STP 1271



# *Science and Technology of Building Seals, Sealants, Glazing, and Waterproofing: Fifth Volume*

*Michael A. Lacasse, Editor*

ASTM Publication Code Number (PCN):
04-012710-10

ASTM
100 Barr Harbor Drive
West Conshohocken, PA 19428-2959
Printed in the U.S.A.

**ISBN: 0-8031-2049-4**
**PCN: 04-012710-10**
**ISSN: 1062-967X**



**Peer Review Policy**

Each paper published in this volume was evaluated by three peer reviewers. The authors addressed all of the reviewers' comments to the satisfaction of both the technical editor(s) and the ASTM Committee on Publications.

To make technical information available as quickly as possible, the peer-reviewed papers in this publication were prepared "camera-ready" as submitted by the authors.

The quality of the papers in this publication reflects not only the obvious efforts of the authors and the technical editor(s), but also the work of these peer reviewers. The ASTM Committee on Publications acknowledges with appreciation their dedication and contribution to time and effort on behalf of ASTM.

Printed in Ann Arbor, MI
June 1996

# Foreword

This publication, *Science and Technology of Building Seals, Sealants, Glazing, and Waterproofing: Fifth Volume,* contains papers presented at the Charles J. Parise Fifth Symposium on Science and Technology of Building Seals and Sealants held in Phoenix, Arizona on 25–26 Jan. 1995. The symposium was sponsored by ASTM Committee C-24 on Building Seals and Sealants. Michael A. Lacasse of the Institute for Research in Construction, National Research Council Canada in Ottawa, presided as symposium chairman and is the editor of this publication.

# Related ASTM Publications

*Building Seals and Sealants, ASTM STP 606,* 1976.
*Science and Technology of Glazing Systems, ASTM STP 1054,* 1989.
*Building Sealants: Materials, Properties, and Performance, ASTM STP 1069,* 1990.
*Science and Technology of Building Seals, Sealants, Glazing, and Waterproofing: Second Volume, ASTM STP 1200,* 1992.
*Science and Technology of Building Seals, Sealants, Glazing, and Waterproofing: Third Volume, ASTM STP 1254,* 1994.
*Science and Technology of Building Seals, Sealants, Glazing, and Waterproofing: Fourth Volume, ASTM STP 1243,* 1995.

# Contents

# Overview

## Introduction

ASTM Committee C24 on Building Seals and Sealants is faced with having to maintain, refurbish, and develop new standards for the sealant, glazing, and gasket industry. The significant efforts made by many within this group over nearly the past four decades have resulted in these standards being used not only within the North American context but also throughout the world. Indeed, these standards very often have formed the basis for the development of national standards in countries of emerging economic importance. Hence, if these standards are to remain relevant in the global marketplace, the committee must keep abreast of new developments and emerging technologies in their domain. This represents a considerable challenge since it is well known that marketplace globilization is creating an overwhelming need to reevaluate the usefulness of existing standards and to bring these to a level that will maintain an industry's competitive edge. The challenges that confront Committee C24 can be met only if this group has the necessary scientific and technological information to assist them in their efforts.

These symposia help fill the need for useful scientific and technological reference and hence are a vital support to the needs of the various task groups within Committee C24. The papers published in this volume were presented during the fifth symposium on the Science and Technology of Seals, Sealants, Glazing, and Waterproofing held 25–26 Jan. 1995 in Phoenix, Arizona. The current series of symposia have been held yearly since 1989, although other proceedings of related symposia have been published prior to this date. A list of related ASTM publications is provided in the front of the book. It is noteworthy to mention that proceedings of the 6th symposium, edited by symposium chairman James C. Myers, will soon be published, and Committee C24 has planned 7th and 8th symposia to be held in the coming years.

The current proceedings provide reviews of technological advances and the most recent developments in the domain of significant importance to both researchers and practitioners in the sealant industry. This would include, for instance, those involved in test method and product development research and practitioners having expertise in the areas of building joint seal and waterproofing design, specification, application, failure diagnosis, maintenance and repair, and other related domains.

This volume contains eighteen contributions covering a wide spectrum of areas related to the science and technology of sealants and waterproofing and cognate systems. The papers have been organized into six categories: joint design and sealant selection; joint sealant failures; design, evaluation, and application of structural silicone glazing sealants; aesthetic performance; recent advances in adhesion testing; and studies related to the long-term performance and aging of sealants and gaskets. Provided below is an overview of the various papers as outlined in the sections above. Concluding remarks will be provided following suggestions for future developments in the domain.

## Joint Design and Sealant Selection

Sealant products in themselves are, typically, reasonably resistant to weathering; they are significantly less durable if inadequately considered in design, improperly specified, or inappropriately installed. Furthermore, the nefarious effects emanating from their premature failure may, in certain instances, bring about the need for very costly repairs to the building envelope or interior spaces, not to mention the very significant cost of removing and reinstalling sealant product in cladding joints. Consequently, the importance of adequate design and proper selection of seal products cannot be overemphasized.

Of the three papers in this section, the first truly represents the state-of-the-art in building joint sealant design. "Recent Developments in the Design of Sealant Joints for Movement and Other Effects" is the most comprehensive design guide and specifiers aid yet to be published and will undoubtedly become the key reference in such matters. It includes information not available in a previous paper by the same author (ASTM STP 1069) and updates information from it. It is a highly recommended paper for all those wanting a thorough review of factors to consider in joint design, and it also provides complete worked examples of a number of joint designs typically found in the field. The following paper, "A Comparison of Different Sealant Butt Joint Shapes," compares the relative performance of three types of sealant butt joints, i.e., square, rectangular, and "hourglass." It is very useful for obtaining detail in regard to the performance of different joint configurations, and the work is supported from the use of finite element analytical techniques and results of mechanical testing of sealant products. "Building Sealants in Russia" gives information on the technical requirements and physical properties of various sealants produced in Russia. It is of interest from the point of view of comparing the level at which the Russian sealant industry operates or has operated in the recent past in relation to that in Europe or North America.

## Joint Sealant Failures: Case Studies and Methods of Evaluation

As noted above, premature failure of joint sealants can bring about very costly decisions for the building maintenance manager, and recourse is often made to obtain the services of a knowledgeable building sealant practitioner competent in joint inspection and seal failure diagnosis. Kenney and Piper are evidently two such practitioners, and their paper, "Case Study on Sealant Reversion," is particularly significant in that it is the first to document sealant reversion in the field. Of importance as well are the techniques used in determining the possible cause of premature failure. Kenny et al. use a Fourrier transform infrared (IR) spectroscopic analysis of the material to determine its peak intensity at a given wavenumber. This provides the necessary technical basis to understand this somewhat obscure phenomena that has, therefore, not been adequately explained.

"History of Building Joint Sealants" is not a history of the technological developments in the sealant industry per se. Rather, it is aimed at investigators trying to identify the types of sealants used in either older buildings or when a building was last refurbished. A review of methods useful in investigating premature sealant failures is given and an understanding is provided of the significance of when different products came onto the market with regard to failure investigations.

## Structural Silicone Glazing: Design, Evaluations, and Applications

The rapidly expanding use of structural sealant glazing (SSG) systems in the past decade and a half has necessitated the development of standards for their use that take into consideration not only their mechanical properties but also their time-dependent performance attributes. The paper by Gutowski et al., "Structural Silicone in Curtain Walls," represents a significant contribution to this area and offers an understanding of proper design based on extensive laboratory studies related to the durability of these materials.

That SSG materials have so far met with appreciable success is not in question; that they can withstand continued aging over several decades in harsh environments has yet to be proven conclusively. The treatment offered in "Physical-Chemical Changes in Silicone Elastomers Used for Building Sealants" sheds some light on the potential changes that can occur in these materials under specific environmental conditions.

It is also apparent that these materials can be used for a number of different types of glazing systems. The paper "Structural Behavior of Silicone Bonded Glass Block Panels" provides an example of a research project aimed at demonstrating the viability of novel uses of structural silicone sealant.

## Aesthetic Performance of Sealants

In this section, three papers are provided on the aesthetic performance of sealants. It should be noted that this topic was the focus of abundant discussion during the symposium presentations; hence, the subject cannot be considered trivial. In fact, many buidling owners, and property managers as well, are especially keen on maintaining the aesthetic appeal of their building facades. Hence, products must be chosen that can maintain an adequate level of performance with the minimum of maintenance over extended periods of time.

Such issues have been thoroughly dealt with by O'Neil and Wolf in a previous offering by the same authors (published in ASTM STP 1243), and the current paper builds on this aforementioned effort in significant detail. "Effects of Weatherproofing Sealants on Building Aesthetics—Part II," together with Part I, can collectively be taken as exceptionally useful reference documents on this topic. Practitioners wishing to avail themselves of information on the types of tests and useful standards in this area should refer to these manuscripts.

"Staining Potential of Sealants in/on Exterior Wall Substrates" is a contribution reporting on the results of a number of field investigations of exterior walls of high-rise buildings and the subsequent laboratory work on various commercially available sealants. It provides insights into why certain sealants have a propensity to stain particular substrates, and it is useful to designers who wish to make appropriate choices of materials in cases where staining may potentially be a problem.

Following this theme is a laboratory study by Farmer and Cechner entitled: "Laboratory Testing of Sealants with a Marble Substrate." This work was undertaken on three polyurethane sealants, and an evaluation is given of the effects on porous substrates of plasticizer migration and the adhesion of these products.

## Advances in Adhesion Testing and Adhesion Promoters

Substantial efforts have been made in recent years to enhance adhesion testing since it has been shown, as suggested by Cerra and Gutowski, that existing adhesion tests for sealants lack the discriminatory ability necessary to determine the relative adhesion performance of different products to a given substrate. In this area, "Performance-Based Adhesion Testing of Structural Sealants" makes important advances in the domain. This paper reports on the evaluation of a test method designed to degrade the adhesive interface at stress levels below that of the cohesive strength of the product. Using this method, adhesive failure is induced that can result in a true measure of adhesive performance of the product.

Another fundamentally sound research effort is provided in "An Analysis of the 180° Peel Test for Measuring Sealant Adhesion" by Shephard and Wightman. This work demonstrates that no single adhesion-in-peel measurement adequately describes the integrity of the sealant joint because this type of test is specific to the test, specimen geometry, and rate of peel. Alternative methods are suggested that yield useful sealant/substrate adhesion properties.

Hubin-Eschger describes a development project of a novel sealant curing system in "Improvement of Initial Adhesion on Aluminum to Insulating Glass Polyurethane Sealants Using A Tertiary Amine Based Curing System." This curing system, developed for a polyurethane-based sealant used for insulated glass units, provides the product with the necessary mechanical and permeance properties to achieve an adequate seal. It is evaluated for enhanced performance in initial adhesion to aluminum in comparison to an existing but potentially environmentally harmful product.

## Studies Related to the Long-Term Performance and Aging of Sealants and Gaskets

The final section is comprised of four papers treating in general the long-term performance of sealant products. The first two, in particular, are concerned with the use of automated cyclic testing devices for inducing fatigue failure in sealants. The paper "The Effect of Artificial Weathering and Movement Accommodation Testing on Building Sealants" by Beasley and Jenkins reports on the evaluation of seven types of sealant subjected to a cyclic test regime after aging in artificial weathering conditions. Commentary is offered on the effect of the aging procedure and its impact on the survival of specimens to cyclic testing. The work by Lacasse et al., "Evaluation of Cyclic Fatigue as a Means of Assessing the Performance of Construction Joint Sealants, Polyurethane Sealants," elaborates on the aging of polyurethane sealant products using cyclic fatigue testing and adds to their previous work (ASTM STP 1243) evaluating silicone sealants using the same methods.

There is continued interest in reviewing performance tests of products subjected to standard test methods; this has been adequately dealt with by Fiorillo in "Evaluation of Various Generic Types of Building Sealants Against ASTM C920—Standard Specification for Elastomeric Joint Sealants," in which results are presented on the testing of ten high-performance sealant products.

"A Review of Recent Developments and Improvements in Accelerated Outdoor and Indoor Xenon Weathering Devices and Methodologies" by Putman et al. provides some insight into accelerated outdoor weathering and artificial indoor weathering techniques with emphasis on recent improvements in the technology that have permitted developing new test methods in this area. This is by no means an exhaustive review of the subject as these have been more thoroughly dealt with elsewhere (see, e.g., Brown, R. P., *Polymer Testing,* Vol. 12, 1993, pp. 459–466; Kockott, D., *Polymer Degradation and Stability,* Vol. 25, 1989, pp. 181–208). However, it does provide the industrialists' viewpoint offered by a group knowledgeable in the field.

## Future Work

There are a number of standards and guides the committee may develop in the future, and the broad areas of development, outlined below, are provided principally for the benefit of those not as well versed in the domain as members of ASTM Committee C24. To members, these items are generally well known, and those who are knowledgeable in this area would, doubtless, be able to provide more depth and detail to that which is suggested in the outline. Nonetheless, the listing of these future endeavours has merit from the point of view of fostering useful discussion. They have been classified in two distinct areas: materials and methods.

*Materials*

- Increased use of environmentally friendly ("green") products implying a move from:
  - ⇒ solvent-based products to water-borne products;
  - ⇒ solids-based and metal-catalyzed products to other systems;
  - ⇒ increased use of thermoplastic elastomers and other recyclable products.
- Increased use of gasket technology for sealing building envelopes.
- Development of completely designed jointing systems integral to the building envelope system.

*Methods*

- *Design and Installation:* use of information technologies for accessing information and optimizing the process of design, selection, specification, and application techniques specific to the type of cladding being designed.
- *Maintenance:* increased implementation of maintenance planning, methods and techniques of inspection, inspection strategies, and improved failure diagnostic techniques.
- *Testing and Evaluation:* increased use of analytical techniques to assess the performance of sealant products, of small sample testing, and the development of probabilistic methods for predicting long-term performance of products in the laboratory and the field.

The above represent but a short list of future developments in the domain without offering pertinent details, as this would be a report in itself. It is hoped, however, that in making these brief contributions, they will find use to those active in sealant standards technical committee work of ASTM Committee C24 and elsewhere, thereby stimulating discussion on these topics.

In closing, I would like to gratefully acknowledge the many generous contributions submitted by the authors, as well as the dedicated efforts of reviewers and ASTM personnel who helped make this publication possible. Thanks are also extended to the COP representative for these proceedings, George Beestman, whose guidance and thoughtful advice were particularly useful in the review and preparation of this text.

*Michael A. Lacasse*

Symposium chairman and editor,
National Research Council Canada,
Institute for Research in Construction,
Materials Laboratory,
Ottawa, Ontario,
Canada

# Joint Design and Sealant Selection

Thomas F. O'Connor[1]

# RECENT DEVELOPMENTS IN THE DESIGN OF SEALANT JOINTS FOR MOVEMENT AND OTHER EFFECTS

---

**REFERENCE:** O'Connor, T. F., **"Recent Developments in the Design of Sealant Joints for Movement and Other Effects,"** Science and Technology of Building Seals, Sealants, Glazing, and Waterproofing: Fifth Volume, ASTM STP 1271, Michael A. Lacasse, Ed., American Society for Testing and Materials, 1996.

**ABSTRACT:** Often, thermal movement and other performance factors are inadequately considered in the design of sealant joints. Performance factors that affect sealant joint design are described, including: thermal movement; moisture movement; live and dead loads; creep, elastic frame shortening, and shrinkage of concrete structures; and material, fabrication, and erection tolerances. Three calculation examples will be described for joints in new construction that accommodate movement: (1) an expansion joint in a masonry wall, (2) a joint between different building systems, and (3) a floor line expansion joint for a multi-story curtain wall and two examples for joints in remedial work: (4) a fillet (triangular) joint between a curtain wall and masonry wall and (5) a bridge joint for a precast concrete panel system..

**KEYWORDS:** bridge joint, butt joint, construction tolerance, expansion joint, fillet joint, moisture movement, performance factor, sealant joint design, thermal movement

---

## NOMENCLATURE

$\alpha_a$ Aluminum linear coefficient of thermal movement
$\alpha_b$ Brick linear coefficient of thermal movement

---

[1]Vice President and Director of Building Technology, Smith, Hinchman & Grylls Associates, Inc., 150 West Jefferson Avenue, Suite 100, Detroit, MI 48226, USA.

$\alpha_c$ Concrete linear coefficient of thermal movement
$\alpha_x$ Particular material linear coefficient of thermal movement
$A$ Solar absorption coefficient
$A_b$ Brick solar absorption coefficient
$A_c$ Painted aluminum solar absorption coefficient
$B$ Sealant backing length
$C_b$ Brick masonry construction tolerance
$C_c$ Curtain wall construction tolerance
$C_e$ Erection tolerance
$C_f$ Fabrication tolerance
$H$ Heat capacity constant
$I$ Moisture induced irreversible movement
$J_c$ Joint compression
$J_e$ Joint extension
$J_l$ Joint longitudinal extension
$J_t$ Joint transverse extension
$L$ Sealant joint spacing
$\Delta L_{bh}$ Dimensional change due to brick horizontal thermal movement
$\Delta L_{bv}$ Dimensional change due to brick vertical thermal movement
$\Delta L_{cv}$ Dimensional change due to curtain wall vertical thermal movement
$\Delta L_i$ Dimensional change due to irreversible moisture movement
$\Delta L_{ih}$ Dimensional change due to horizontal irreversible moisture movement
$\Delta L_{iv}$ Dimensional change due to vertical irreversible moisture movement
$\Delta L_s$ Dimensional change due to summer installation thermal movement
$\Delta L_w$ Dimensional change due to winter installation thermal movement
$\Delta L_x$ Dimensional change for a particular condition
$M_c$ Movement due to spandrel beam creep deflection
$M_l$ Movement due to live load
$M_w$ Movement due to wind load
$R$ Moisture induced reversible movement
$S$ Sealant movement capacity
$T_a$ Hottest summer air temperature
$T_i$ Winter interior air temperature
$T_{is}$ Maximum summer installation wall surface temperature
$T_{iw}$ Minimum winter installation wall surface temperature
$T_s$ Hottest summer wall surface temperature
$T_w$ Coldest winter wall surface temperature
$\Delta T_m$ Maximum expected temperature difference
$\Delta T_s$ Summer installation temperature difference
$\Delta T_w$ Winter installation temperature difference
$\Delta T_x$ Temperature difference for a particular installation time
$W$ Final designed joint width
$W_e$ Joint width after elongation
$W_h$ Joint width required for horizontal movement
$W_v$ Joint width required for vertical movement
$W_m$ Joint width required for movement

$W_w$ Joint width required for wind induced movement

## INTRODUCTION

Architects and other design professionals, for aesthetic reasons, have desired to limit the width of sealant joints on the exterior walls of new buildings. Frequently the admonition has been "I want to see a 6 mm (1/4 in) wide joint". Analysis of the performance factors that affect a sealant joint is necessary to determine if a 6 mm (1/4 in) wide joint will be effective in maintaining a durable seal [1] against the passage of air and water. If the performance factors are not understood, and included in the design of the joint [1], then failure is a distinct possibility.

Failure can result in increased building energy usage, water infiltration, and deterioration of building systems and materials. Infiltrating water can cause spalling [1] of friable building materials such as concrete, brick, and stone; corrosion of ferrous metals; and decomposition of organic materials, among other effects. Deterioration is often difficult and very costly to repair, with the cost of repair work usually greatly exceeding the original cost of the sealant joint work.

The objectives of this paper are to: (1) describe some of the performance factors that are normally considered in sealant joint design, and (2) provide sample calculations for sealant joint design. The selection and properties [2, 3] and installation [4] of sealants will not be described. This paper revises previously published papers [5, 6] and also provides updated information for fillet and bridge type sealant joints not previously available.

## PERFORMANCE FACTORS

The following describes some of the performance factors that can influence sealant joint design either individually or in various combinations.

### Material Anchorage

The type and location of various wall anchors has an impact on the performance of a sealant joint. Large precast concrete panels with fixed and moving anchors, brick masonry relieving angle deflection, and metal and glass curtain wall fixed and moving anchorages are examples of anchorage conditions that must be evaluated when designing sealant joints for movement. These anchors have an effect on the length of wall material or the deflection characteristics to be included in the design of the sealant joint width.

### Thermal Movement

Walls of buildings respond to ambient temperature change and solar radiation by either increasing or decreasing in volume and therefore in linear dimension. The dimensional change of wall materials causes a change in the width of a sealant joint opening, producing a movement in the sealant.

For information, thermal movement may need to be evaluated at different stages in a building's life; for example, expected temperature differentials may need to be considered for the building when it is (1) under construction, (2) unoccupied and unconditioned, and (3) occupied and conditioned. Each of these stages will have different interior environmental conditions, and depending on the building system being analyzed for movement, one of those stages may produce the maximum expected thermal movement. The joint opening width, depending on construction procedures and wall system types, could be established during one of the stages.

Determining realistic wall or material surface temperatures, to establish the expected degree of movement, can be challenging. The author uses reference [7] which lists winter and summer design dry bulb air temperatures for many cities, to assist in establishing surface temperatures for use in joint width calculations.

The coldest winter wall surface temperature, $T_w$, is established using the winter design dry bulb air temperature. This is realistic since the exterior wall surface will be within a few degrees of the air temperature, depending on the degree of insulation of the wall and a lack of solar radiation. Reference [7] lists the 99 and 97.5 percent design dry bulb temperatures. Using the 99 percent value, which represents a total of 22 hours during the winter at or below the listed temperature, seems appropriate. Steady state heat flow calculations to determine the surface temperature does not seem necessary for most uses.

The hottest summer wall surface temperature, $T_s$, is established as follows. The starting point is Reference [7] which lists summer 1, 2.5, and 5 percent design dry bulb air temperatures. Using the 1 percent value to establish the hottest summer air temperature, $T_a$, which represents a total of 30 hours during the summer at or above the listed temperature, seems appropriate.

Walls also receive solar radiation which warms the wall surface. Wall surface radiative heat gain, A(H), must be included with $T_a$ when establishing $T_s$, which is found using Equation 1 [8, 9].

$$T_s = T_a + A(H) \qquad (1)$$

The heat capacity constant H, shown in both S.I. and (inch-pound units), should be used with the following guidelines. Low heat storage capacity materials or walls are represented by well insulated metal panel curtain walls; and high heat storage capacity materials or walls are represented by precast concrete or brick masonry. If light colored or other reflective adjacent surfaces reflect solar radiation to the wall surface, then constants that include this effect should be used in lieu of the previous constants [9]. Table 1 lists values for H to use in Equation 1.

TABLE 1--Heat capacity constants.

| Condition | Constant, H | Condition | Constant, H |
|---|---|---|---|
| Low | 56 (100) | High | 42 (75) |
| Low, with reflection | 72 (130) | High, with reflection | 56 (100) |

Equation 1 is for Easterly, Westerly, and Southerly facing walls that receive direct solar radiation. Even Northerly facing walls receive some direct solar radiation during the summer, either in the morning or late in the afternoon. During the day, when there is no direct solar radiation on a Northerly facing wall, there is an indirect solar warming effect caused by diffuse radiation reflected onto the wall from particles suspended in the atmosphere. The increase to the hottest summer air temperature due to this effect is relatively small, probably in the range of 6 to 17°C (10 to 30°F), depending on locale, orientation, wall materials, and degree of insulation. In most cases, the sealant joint widths of buildings are established on the basis of the wall with greatest solar exposure, with that joint width repeated for all the building walls. Hence, it is usually not necessary to calculate a solar radiation component for a Northerly facing wall.

TABLE 2--Solar absorption coefficients.

| Material | Coefficient, A | Material | Coefficient, A |
|---|---|---|---|
| Aluminum, clear finish | 0.60 | Glass, clear, 6mm (1/4 in) | 0.15 |
| Aluminum paint | 0.40 | Glass, tinted, 6mm (1/4 in) | 0.48 - 0.53 |
| Aluminum, dull | 0.40 - 0.65 | Glass, reflective, 6mm (1/4 in) | 0.60 - 0.83 |
| Black, nonmetallic | 0.85 - 0.98 | Marble, White | 0.58 |
| Asphalt, slate | | Paint | |
| Mineral Board, uncolored | 0.75 | Dark red, brown or green | 0.65 - 0.85 |
| Mineral Board, white | 0.61 | Black | 0.85 - 0.98 |
| Brick, light buff (yellow) | 0.50 - 0.70 | White | 0.23 - 0.49 |
| Brick, red color | 0.65 - 0.85 | Surface Color, Black | 0.95 |
| Brick, white | 0.25 - 0.50 | Surface Color, Dark Grey | 0.80 |
| Concrete, uncolored | 0.65 | Surface Color, Light Grey | 0.65 |
| Copper, tarnished | 0.80 | Surface Color, White | 0.45 |
| Copper, patina | 0.65 | Tinned Surface | 0.05 |
| Galvanized Steel, | 0.40 - 0.65 | Wood, Smooth | 0.78 |
| Galvanized Steel, white | 0.26 | | |

Table 2 lists some values for the solar absorption coefficient, A, to use in Equation 1 [8, 9, 10]. Reference [11] can be used in lieu of the values in Table 2 and Equation 1 to determine various glass product surface temperatures. The solar absorption coefficients may need to be modified based on past experience and an evaluation and understanding of the wall construction. Judgement should still be used by the design professional in establishing these values.

In addition to the temperature extremes the wall material will experience, its linear coefficient of thermal movement ,$\alpha$, must also be determined. The amount of linear thermal movement is usually expressed, for each construction material, as an increase or decrease per unit length per degree change in material temperature. Reference [4] includes a table that lists the linear coefficient of thermal movement for many of the commonly used construction materials. For a wall or panel construction that is a composite of materials, an appropriate coefficient of thermal movement should be determined for the composite assembly.

## Moisture Induced Movement

Some materials respond to changes in their water or water vapor content by increasing in dimension when water content is high and decreasing in dimension when water content is low [12]. Materials susceptible to this reversible effect include: wood, some natural building stones, concrete, face brick, and concrete block. Some materials also experience an irreversible change in dimension with the passage of time.

TABLE 3 -- Moisture induced movement.

| Material | Movement, Percent | |
|---|---|---|
| | Reversible R | Irreversible I |
| Concrete, gravel aggregate | 0.02 - 0.06 | 0.03 - 0.08 (-) |
| Concrete, limestone aggregate | 0.02 - 0.03 | 0.03 - 0.04 (-) |
| Concrete, lightweight aggregate | 0.03 - 0.06 | 0.03 - 0.09 (-) |
| Concrete block, dense aggregate | 0.02 - 0.04 | 0.02 - 0.06 (-) |
| Concrete block, lightweight agg. | 0.03 - 0.06 | 0.02 - 0.06 (-) |
| Face brick, clay | 0.02 | 0.02 - 0.07 (+) |
| Limestone | 0.01 | NA |
| Sandstone | 0.07 | NA |

(-) indicates a reduction, (+) indicates an increase, and NA not available.

Provided in Table 3 [13] are values for moisture induced reversible movement, R, as well as irreversible movement, I, for various types of materials. Reversible movement is based on the likely extremes of in-service moisture content and irreversible movement on the period from material manufacture to its maturity. The use of steel reinforcement will usually lessen the Table 3 concrete values. For "Face brick, clay" the reference [14] specific value of 0.03 can be used for I in lieu of the range of values, if appropriate. In general, cement-based products decrease in dimension and fired clay products increase in dimension irreversibly as they equilibrate with the environment.

For sealant joints, the dominant effect on a reversible change in joint width is usually due to temperature change of a wall surface material. The inclusion of reversible moisture induced movement with thermal movement may not be a truly additive effect. Moisture content tends to decrease with a rise in wall surface temperature and increase with a drop in wall surface temperature, thereby producing thermal and moisture induced movements that are somewhat compensating but that may not necessarily occur simultaneously. The net sealant joint movement due to thermal and moisture effects may be difficult or impossible to determine, so some judgement must be used by the design professional when reversible moisture induced movement is considered.

## Live Load Movement

Deflection caused by structure or floor live loading should be considered for horizontal sealant joint openings; as is done for example, in designing joints for multi-story construction. The building structural engineer can supply live load deflection criteria for sealant joint design.

Actual live loads can be highly variable [15]. A multi-story building, with the same design live load for all floors, will have the actual live load (which can be substantially less than a code prescribed value) vary from floor-to-floor and from one area of a floor to another. Very rarely will the live load be uniform everywhere. Where live load and thus deflection of a structure varies, the relative difference in live load deflection between floors should be considered in joint width design.

Most often live load deflection occurs after the joint has been sealed and, therefore, could be considered an irreversible narrowing of the sealant joint opening, provided the loading conditions remain relatively static. If live loading will be highly variable, such as in a warehouse, then live load deflection could be treated as a reversible movement. The design professional should evaluate these situations and determine how live load deflection is best accommodated.

## Dead Load Movement

Deflection caused by structure or floor dead loading should also be considered for horizontal sealant joint openings. The building structural engineer can supply dead load deflection criteria for sealant joint design.

Dead load deflection of a structure usually occurs before a joint is sealed. There may be a portion that could occur after a joint has been sealed; for instance, when fixed equipment may be installed. Dead load deflection is an irreversible narrowing of the sealant joint opening width for most applications.

Wind Load Movement

Depending on building type, framing system, and anticipated wind load, lateral sway or drift of a building and its effect on sealant joints in a wall may have to be considered. The per-story lateral sway or drift can be determined by the building structural engineer. Lateral sway or drift can occur both normal to and in the plane of the wall and both effects on a sealant joint will need to be considered.

Elastic Frame Shortening

Multi-story concrete structures, and to a lessor degree steel, shorten elastically almost immediately due to the application of loads [15, 16]. Frame shortening, the degree of which can be estimated by a structural engineer, will cause an irreversible narrowing of horizontal sealant joint openings in multi-story construction. Frame shortening can be compensated for by building each floor level slightly higher, in effect negating most of the shortening, or the narrowing of joint width can become another performance factor considered in the design of a sealant joint.

Some of the frame shortening effect will occur before the wall cladding is erected and the size of the joint opening is established. Presently, it is common practice to determine the amount of shortening that occurs before the joint opening is established using an informed and conservative estimate.

Creep

The time dependent deformation of materials while loaded, in particular a concrete structure, should be included in sealant joint design. This deformation, which occurs at a decreasing rate as time progresses, can cause a continuing decrease in the width of horizontal joint openings in multi-story and other buildings. Creep, in contrast to elastic frame shortening, can occur over a long period of time [15, 16]. The building structural engineer can provide creep deflection criteria for sealant joint design.

Shrinkage

Concrete framed structures will undergo long-term shrinkage for a period of months [15, 16]. Shrinkage is mainly due to the initial drying of the concrete mix water. The rate of shrinkage is dependent on the amount of water present, ambient temperatures, rate of air movement, relative humidity of the surrounding air, the shape and size of the concrete section, and the amount and type of aggregate in the concrete mix, among others. Table 2 and Reference [13] list guidelines for some shrinkage values for concrete and other materials. Shrinkage criteria can be provided by a structural engineer and included in sealant joint design or the shrinkage effect can be compensated for by building each floor level slightly higher. In any event, shrinkage effects should be included in the design of horizontal joints in multi-story construction.

Some of the frame shrinkage effect will occur before the wall cladding is erected and the size of the joint opening is established. Presently, the amount of shrinkage that occurs before the joint opening is established is determined by an informed estimate and, therefore, should be conservative.

Construction Tolerances

ASTM and industry trade associations, among others, establish industry recognized standards for construction tolerances. For some materials or systems there are no industry recognized tolerances or the available tolerances are not directly applicable to sealant joint design. In these instances, the design professional should evaluate the conditions and establish tolerances for the sealant joint work. Tolerances should be indicated on the contract documents for the building, since they establish a level of quality and may affect the cost and performance of the work.

Material--Construction materials have a permissible variation for their dimensions. For example, a face brick is nominally 57 mm (2-1/4 in) high by 203 mm (8 in) long by 89 mm (3-1/2 in) thick. Depending on the type of brick, the permissible manufacturing tolerance or variation could be as much as 4 to 6 mm (5/32 to 1/4 in) for the 203 mm (8 in) dimension, as established by Reference [17]. Material dimensional variation may have to be included as a performance factor in the design of a sealant joint. If material tolerance is not considered, a deficiently sized joint width could result.

Fabrication--Fabricated materials or assemblies of materials also have dimensional variance. Usually factory fabrication will permit a smaller variation in dimension than job site fabrication. For example, factory fabricated unitized metal and glass curtain wall frames may permit ± 2 mm (1/16 in) tolerance or less for the length and width of the panels, while job site assembly of a face brick wall, depending on brick type, may permit no better than ± 6 mm (1/4 in) variance for the constructed width of an expansion joint opening in a wall. Also, a precast concrete wall panel that is 9.1 m (30 ft) long, according to Reference [18] can have a tolerance for that dimension at the time of casting of +3, -6 mm (+1/8, -1/4 in). Realistic fabrication tolerances should be established and enforced so that the designed joint opening width is attained and sealant performance is not compromised, especially by a joint opening that is constructed too narrow in width.

Erection--Frequently, wall materials or systems cannot be placed on a building exactly where called for by the contract documents. Some location variance for building components should be provided so that a deficient joint opening width does not occur. For example, a unitized metal and glass curtain wall panel may be erected no closer then ± 3 mm (1/8 in) to height or lateral locations shown by the contract documents. The above described precast concrete wall panel, again according to Reference [18], may not be able to be placed any closer to its theoretical location than ± 6 mm (± 1/4 in). This erection variation of materials or systems will affect the constructed width of sealant joint openings that occur between wall panels. Erection tolerances must be intelligently developed so that they are realistic and also attainable at the building site.

Basic Joint Movements

There are four basic movements that sealant joints experience. These movements are: compression, extension, longitudinal extension, and transverse extension as illustrated by Fig. 1. Longitudinal and transverse extension produce a shearing effect on the sealant joint as illustrated by Fig. 2. Thermal movement is usually the largest contributor; however, other performance factors can contribute to producing these movements. The following describes these movements.

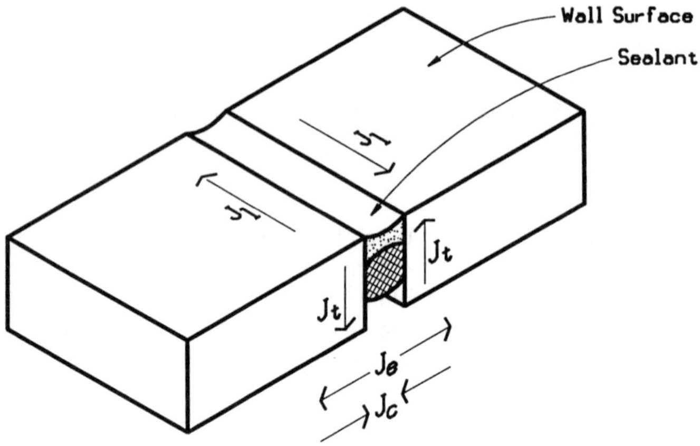


$J_c$ = Compression
$J_e$ = Extension
$J_l$ = Longitudinal Extension
$J_t$ = Transverse Extension

FIG. 1--Typical sealant joint movements.

Compression--A sealant joint that predominantly experiences compression (a narrowing of the opening width), is typically one where the sealant is installed during the cool or cold months of the year. Therefore, when the warm summer months occur the thermal growth of adjacent materials causes a narrowing of the sealant joint opening, thereby compressing the sealant.

Extension--A sealant joint that predominantly experiences extension (an increase in the opening width), is typically one where the sealant is installed during the warm months of the year. Therefore, when the cool or cold months occur the thermal contraction of adjacent materials causes a widening of the joint opening, thereby extending the sealant.

Extension and compression--A sealant joint installed during the fall or spring months, or when temperatures are moderate, can experience both compression as well as extension since the sealant is not installed at or near the hottest or coldest design temperatures. This results in compression during the summer months and extension during the winter months.

Longitudinal extension--A sealant joint that experiences longitudinal extension (a lengthwise shearing type of displacement of one side of the joint opening relative to the other), is typically one that has different materials or systems forming the sides of the joint; for example, a brick masonry wall for one side and an aluminum curtain wall mullion for the other. These materials have different responses to the same temperature change (the brick will change dimension less than the aluminum), resulting in a diagonal lengthening of the sealant due to the differential movement between the materials in a direction parallel to the joint opening. This movement, illustrated by Fig. 2, is dependent on the location of each materials' supports, the unrestrained length of the respective materials, and the expected temperature change for each material. This movement usually reaches its maximum only along part of the length of a sealant joint. Irreversible moisture growth of fired clay products such as face brick can also contribute to the effect.

Transverse extension--A sealant joint that experiences transverse extension (a shearing type of out-of-plane movement, crosswise to the joint face, of one side of the joint opening relative to the other), is typically one that occurs at the juncture of walls that change plane, such as at a corner. As the materials forming the sides of the joint experience thermal movement, a diagonal lengthening of the sealant can occur crosswise to the plane of the sealant joint face as illustrated by Fig. 2.

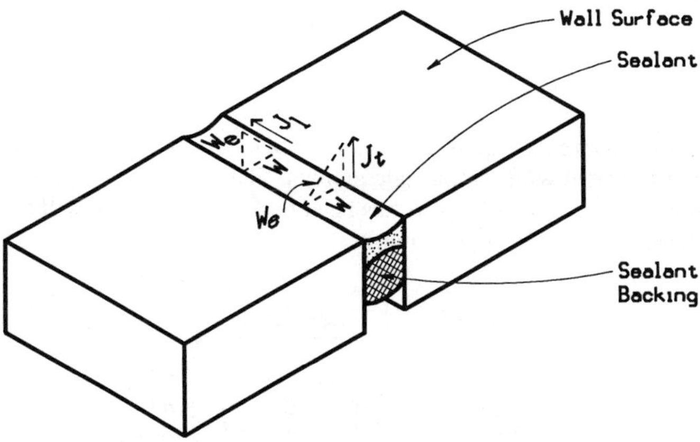


W = Joint Width at Rest
$J_l$ = Longitudinal Extension
$J_t$ = Transverse Extension
$W_e$ = Sealant Elongation, Where $W_e = (1+S)(W)$ and S is the sealant movement capacity expressed as a decimal

FIG. 2--Longitudinal or transverse extension movement.

Movement combinations--Frequently, joints must accommodate more than one of the above described movements. Examples include the previously described extension and compression, as well as extension and/or compression combined with longitudinal or transverse extension. The design professional should evaluate the types of movement the joint will experience and design accordingly.

Sealant stresses in extension and shear--Recently, testing was performed on various polyurethane and silicone sealants [1] to establish how shear stresses produced in those sealants, when in longitudinal or transverse extension, compared with the tensile stresses produced when the sealants were in extension [19]. The results of that work indicate that the use of a trigonometric equation, based on the relationships illustrated by Fig. 2, in conjunction with a sealant's tensile modulus properties, can successfully predict allowable shear movement for a sealant joint, since the shear stress did not exceed the tensile stress for a given elongation.

Interestingly, it was also established that for silicone sealants in general, the stresses produced by shear movement, for a given elongation, were less then those produced by tensile movement for the same elongation. This indicates that for silicone sealants there is the potential of greater movement or elongation capability for sealant joints that experience longitudinal or transverse extension. In general, the author does not take advantage of this extra capacity for silicone sealants, instead opting for a more conservative design or a slightly larger design factor, due to some imprecision in establishing material surface temperatures and other sealant joint performance factors.

## ILLUSTRATIVE EXAMPLES

The following five examples of sealant joints in new and remedial construction (Fig. 3) will be illustrated with sample calculations. The examples are presented in both S.I. and inch-pound units. Rounding-off has been performed so there will not be exact agreement between S.I. and inch-pound units.

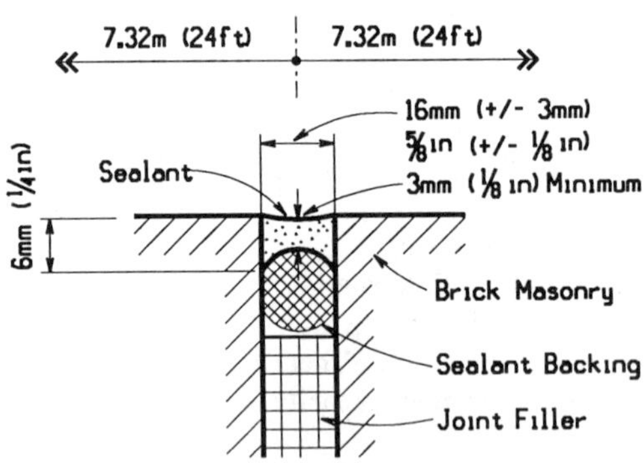


Fig. 3--Butt type sealant joint profile.

Example One: Brick Masonry Expansion Joint

Example description--Brick masonry is used for the exterior walls of buildings because of its durability and aesthetic appearance. However, to maintain durability, the masonry surface should be divided into appropriately sized panels, separated by expansion joints [1], that are designed to respond to thermal movement and other effects. If properly designed expansion joints are not provided they will occur naturally as uncontrolled cracking, with sometimes disastrous results.

For a new building in Detroit, Michigan, the architect has decided that a red brick masonry wall will have expansion joints at 7.32 m (24 ft) maximum spacing. An extruded brick with little dimensional variation will be used which, in conjunction with the detailing of the wall, will permit a tolerance of ± 3.2 mm (1/8 in) for the constructed joint opening width. No masonry construction is expected to occur below 4°C (40°F) in the winter and the maximum wall installation temperature expected during the summer is 35°C (95°F). The above, as well as other performance factors that will effect joint design including the source for the data, are listed in Table 4.

Table 4--Performance factors - Example One.

| Factor | Value | Factor | Value |
|---|---|---|---|
| $L$ | = 7.32 m (24 ft) | A (Table 2) | = 0.65-0.85 |
| $C_b$ | = ± 3.2 mm (±1/8 in) | R (Table 3) | = 0.02 percent |
| $T_w$ [7] | = -16°C (3°F) | I [14] | = 0.03 percent |
| $T_a$ [7] | = 33°C (91°F) | $\alpha_b$ [4] | = 0.0000065 mm/mm/°C (0.0000036 in/in/°F) |
| $T_{iw}$ | = 4°C (40°F) | | |
| $T_{is}$ | = 35°C (95°F) | S | = ± 50 percent |

Temperature differences--The expected brick masonry dimensional change due to thermal movement is determined as follows. No restraining effects of wall ties or joint reinforcement is included in the calculation. The specific solar absorption coefficient for this brick is not known, hence a conservative, value from Table 2 is used. The hottest summer wall surface temperature, $T_s$, is determined using Equation 1, a value for A of 0.85, and a constant for a high heat capacity material taken from Table 1.

$$\begin{aligned} T_s &= T_a + 42A \\ &= 33 + 42(0.85) \\ &= 69^\circ C \end{aligned} \qquad \text{and} \qquad \begin{aligned} T_s &= T_a + 75A \\ &= 91 + 75(0.85) \\ &= 155^\circ F \end{aligned}$$

The coldest winter wall surface temperature, $T_w$, is -16°C (3°F). The maximum expected temperature difference, $\Delta T_m$, is found using Equation 2.

$$\Delta T_m = T_s - T_w \qquad (2)$$

substituting

$$\Delta T_m = 69 - (-16) = 85°C \quad \text{and} \quad \Delta T_m = 155 - 3 = 152°F$$

If the maximum summer and minimum winter installation temperatures, $T_{is}$ and $T_{iw}$, are not known, the above values should be used. Since, for this example, values for $T_{is}$ and $T_{iw}$ are known, then the expected temperature differences, $\Delta T_s$ and $\Delta T_w$ are found using Equations 3 and 4.

$$\Delta T_s = T_{is} - T_w \quad (3) \quad \text{and} \quad \Delta T_w = T_s - T_{iw} \quad (4)$$

substituting, for summer installation temperature difference using Equation 3

$$\Delta T_s = 35 - (-16) = 51°C \quad \text{and} \quad \Delta T_s = 95 - 3 = 92°F$$

and for winter installation temperature difference using Equation 4

$$\Delta T_w = 69 - 4 = 65°C \quad \text{and} \quad \Delta T_w = 155 - 40 = 115°F$$

The above analysis is expressed graphically in Fig. 4.

$T_w$ = -16°C (3°F) $T_{iw}$ = 4°C (40°F) $T_{is}$ = 35°C (95°F) $T_s$ = 69°C (155°F)

|________________|________________|________________|

$\Delta T_w$ = 65°C (115°F)

|------------------------------------------->|

$\Delta T_s$ = 51°C (92°F)

|<-------------------------------------------|

$\Delta T_m$ = 85°C (152°F)

|<-------------------------------------------------------------->|

FIG. 4--Expected thermal movements.

Thermal movements--The change in dimension of a material or system, $\Delta L_x$, due to thermal movement is determined using Equation 5.

$$\Delta L_x = (L)(\Delta T_x)(\alpha_x) \qquad (5)$$

substituting to find $\Delta L_s$, using the summer installation temperature difference $\Delta T_s$ of

51°C (92°F) and $\alpha_b$

$$\Delta L_s = (7.32)(1000)(51)(0.0000065) = 2.4266 \text{ mm}$$

and

$$\Delta L_s = (24)(12)(92)(0.0000036) = 0.0954 \text{ in}$$

and to find $\Delta L_w$, using the winter installation temperature difference $\Delta T_w$ of 65°C (115°F) and $\alpha_b$

$$\Delta L_w = (7.32)(1000)(65)(0.0000065) = 3.0927 \text{ mm}$$

and

$$\Delta L_w = (24)(12)(115)(0.0000036) = 0.1192 \text{ in}$$

In the above and other calculations that follow, the value of 1000 for S.I. and 12 for inch-pound units is a factor to convert meters to millimeters and feet to inches respectively.

Moisture movements--The effect of a material's moisture growth is determined as follows. The reversible moisture content value, R, is based on measurements made from an extreme wet (but not submerged) to an extreme dry external exposure [13]. A review of the technical data for this particular brick indicates a low rate of water absorption and neither an extreme wet nor dry exposure is expected. The reversible moisture growth is expected to be negligible and will not be considered.

For dimensional change due to irreversible moisture growth, I, a range of values is indicated in Table 3 for the face brick. If specific data for a particular brick is unavailable, the values toward the high end of the range could be used. For this example, the value of 0.03 percent recommended in Reference [14] will be used. Some of the dimensional change due to irreversible moisture growth of the brick will occur before the joint opening is built; however, determining this partial value may not be possible and the full value should be used for calculation. The change in dimension, $\Delta L_i$, due to irreversible moisture movement is determined using Equation 6.

$$\Delta L_i = (I/100)(L) \tag{6}$$

substituting

$$\Delta L_i = (0.03/100)(7.32)(1000) = 2.1960 \text{ mm} \quad \text{and} \quad \Delta L_i = (0.03/100)(24)(12) = 0.0864 \text{ in}$$

Where necessary for calculations that include reversible moisture movement, the reversible moisture movement effect, R, is also found using Equation 6; however, substituting reversible movement, R, for irreversible movement, I.

For this example, the expected sealant movement is caused by dimensional changes due to thermal and irreversible moisture movement. Since a period of time will elapse before the brick attains its estimated long term moisture movement (growth), the joint designer, depending on joint configuration and the types of joint movement, may need to evaluate sealant movement both with and without the irreversible moisture movement affect to determine which condition will produce the largest net movement.

Movement analysis--For this example, Fig. 5 indicates that for a winter joint opening installation, since the movements are additive and not compensating, both effects should be included when determining the required joint width. For a summer joint opening installation, the dimensional change due to thermal movement will eventually be reduced by the dimensional change due to the ultimate irreversible movement; therefore, the summer condition should be evaluated without the irreversible movement effect since that condition will result in the greatest joint movement early in the life of the sealant joint.

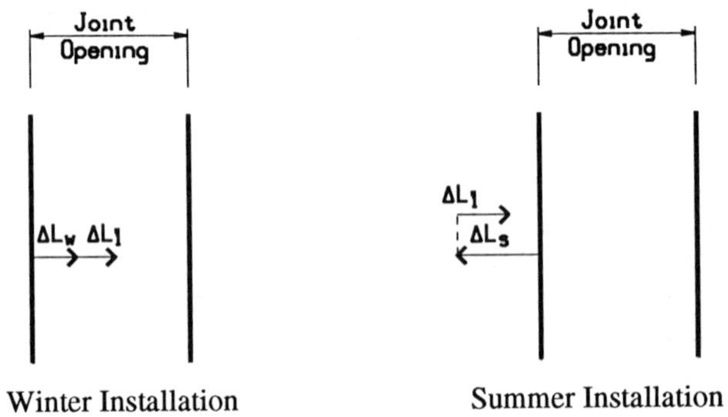


FIG. 5--Movements related to installation time.

Since the thermal movement calculations indicate that dimensional change due to winter installation, $\Delta L_w$, is larger than the dimensional change due to summer installation, $\Delta L_s$, and for this example, dimensional change due to irreversible moisture growth, $\Delta L_i$, is additive to $\Delta L_w$, only the winter installation condition needs to be evaluated since it will produce the largest joint movement.

Additionally, a narrowing of the joint opening is caused by the negative construction tolerance. The required joint width is determined using the movement capacity of the sealant and the previously determined thermal and irreversible moisture movement dimensional changes to calculate a width to satisfy the expected sealant movement. The construction tolerance narrowing effect is then added to that width to arrive at the designed width of the joint opening.

Sealant joint width--A sealant with a ± 50 percent movement capacity will be specified. The author does not believe it is prudent to use a sealant at its manufacturer rated movement capacity. Doing so does not provide any insurance for unknowns and imprecision in establishing surface temperature and other performance factors. Presently, the author evaluates each joint design to determine an appropriate movement capacity to use for calculating the required joint width. For this example, where only extension and compression movement and one material (face brick) are being considered, the sealant will be used at 80 percent of its rated movement capacity to provide insurance for unknowns or error. Therefore, the ± 50 percent sealant will be used at ± 40 percent. Based on this and the above, the required joint width, $W_m$, to satisfy dimensional changes due to thermal movement, $\Delta L_w$, and the irreversible growth effect, $\Delta L_i$, is found using Equation 7.

$$W_m = (\Delta L_w + \Delta L_i)/S \qquad (7)$$

substituting

$W_m = (3.0927 + 2.1960)/0.40 = 13.2218$ mm and $W_m = (0.1192 + 0.0864)/0.40 = 0.5140$ in

The final designed width, W, of the sealant joint, including the negative construction tolerance, $C_b$, is determined using Equation 8.

$$W = W_m + C_b \qquad (8)$$

substituting

$W = 13.2218 + 3.2000 = 16.4218$ mm and $W = 0.5140 + 0.1250 = 0.6390$ in

and rounding off

$$W = 16 \text{ mm } (5/8 \text{ in})$$

The contract documents should indicate the final designed joint width, as illustrated by Fig. 3, with the permissible construction tolerance as follows.

16 mm (± 3 mm) and 5/8 in (± 1/8 in)

Sealant joint depth--The depth of the sealant at its adhesion surface is established as follows. In general, sealant adhesion depth at the substrate [1] surface should be no less than 6 mm (1/4 in) for joint widths from 6 mm (1/4 in) to 12 mm (1/2 in). From 12 mm (1/2 in) to 18 mm (3/4 in) in width the adhesion area should be 1/2 of the joint width. For joint widths in excess of 18 mm (3/4 in) to 50 mm (2 in) a maximum depth of 9 mm (3/8 in) should be considered. For joint widths in excess of 50 mm (2 in) the sealant manufacturer should be consulted. If the ratio of sealant depth to width (the shape factor), which for this example is approximately 1:2.5, is not properly established, the ability of the sealant to accommodate movement can be seriously compromised.

Reference [4] should also be consulted for sealant design and installation guidelines. Those who are interested in a more mathematically detailed approach to establishing masonry surface temperatures and thermally induced movement of brick, should consult reference [10].

Example Two: Joint Between Different Materials or Systems

Example description--Frequently, different materials or building systems abut on the face of buildings with a sealant joint performing as a transition and a seal between the cladding systems. Often, this sealant joint is subject to combinations of thermal movements that occur in more than one direction and differing tolerances of construction from the materials or systems forming the sides of the joint. If improperly designed, longitudinal tearing of the sealant along part of its length is an often seen failure of this joint type.

The masonry wall from Example One terminates at a white painted aluminum mullion and glass curtain wall enclosed lobby. The curtain wall and masonry wall are both 6.1 m (20 ft) high and are separated by a sealant joint. Both wall systems are supported at the bottom on a foundation wall; therefore, all vertical thermal movement occurs at the top of both walls. The masonry wall will be erected before the curtain wall, and the curtain wall could be erected between the winter wall surface temperature ,$T_w$, and the expected maximum summer installation temperature, $T_{is}$. The expected construction tolerances for the sealant joint are ± 4.8 mm (3/16 in) for the location of the end of the brick wall, $C_b$, and since the curtain wall will be erected as a stick system in the field, it is reasonable to expect ± 3.2 mm (1/8 in) for exactness of lateral location of the vertical mullions, $C_c$. The above, as well as other performance factors that will affect joint design, are listed in Table 5. The same sources of data for the previous example are used.

TABLE 5--Performance factors - Example Two.

| Factor | Value | Factor | Value |
|---|---|---|---|
| $S$ = | ± 50 percent | $A_c$ = | 0.45 |
| $L$ = | 6.1 m (20 ft) | $T_{is}$ = | 35°C (95°F) |
| $C_b$ = | ± 3.2 mm (1/8 in) | $I$ = | 0.03 percent |
| $C_c$ = | ± 4.8 mm (3/16 in) | $\alpha_b$ = | 0.0000065 mm/mm/°C (0.0000036 in/in/°F) |
| $T_w$ = | -16°C (3°F) | | |
| $T_s$ = | 33°C (91°F) | $\alpha_a$ = | 0.0000238 mm/mm/°C (0.0000132 in/in/°F) |
| $A_b$ = | 0.65-0.85 | | |

Material or system dimensional changes affecting the sealant joint separating the two systems has vertical and horizontal components. The horizontal component causes a compression and/or extension of the joint while concurrently the vertical component causes longitudinal extension, which increases in magnitude towards the top of the walls. For this example, the horizontal component for the curtain wall will be accommodated within the curtain wall system, while the horizontal component for the brick masonry, $\Delta L_{bh}$, including irreversible moisture growth, $\Delta L_{ih}$, will be accommodated by the sealant joint. The vertical component is the differential movement that occurs between the vertical component for the brick masonry, $\Delta L_{bv}$, and the vertical component of the curtain wall, $\Delta L_{cv}$, due to their differing surface temperatures and coefficients of thermal movement, and irreversible moisture growth, $\Delta L_{iv}$, of the brick masonry. See Fig. 7 for a graphic depiction of the relationship of moisture growth and thermal movement to the time of joint opening installation.

Joint movement effects that include irreversible moisture growth or shrinkage of materials need to be evaluated both with and without the irreversible moisture effect, since the effect occurs gradually over a period of time. When the joint is new it will experience dimensional change from mostly thermal movement. However, as the building ages, brick masonry dimensional change due to irreversible moisture movement increases. The maximum movement the sealant joint will experience may occur either with or without the irreversible moisture movement effect.

Temperature differences--The expected brick masonry wall thermal movement is determined as follows. The wall surface temperatures in winter and summer that were previously determined, $T_w$ = -16°C (3°F) and $T_s$ = 69°C (155°F) respectively, yield a maximum expected temperature difference of $\Delta T_m$ = 85°C (152°F).

Thermal movement of the vertical aluminum curtain wall mullions is determined as follows. The surface temperature extremes are, for winter, $T_w$ = -16°C (3°F) which, for this example, is also the winter installation temperature, $T_{iw}$; and by Equation 1, using a constant for a low heat capacity material, for summer, $T_s$ = 58°C (136°F). From Equation 2, the maximum temperature difference for the curtain wall, $\Delta T_m$, (equivalent to the winter installation temperature difference, $\Delta T_w$) is 74°C (133°F). Thermal movement, based on a maximum curtain wall summer installation temperature $T_{is}$ of 35°C (95°F), should also be evaluated with the curtain wall minimum winter installation temperature $T_{iw}$ of -16°C (3°F) and the maximum curtain wall summer surface temperature of 58°C (136°F). For this example, calculations result in curtain wall summer installation temperature differences, $\Delta T_s$, of 51°C (92°F) and 23°C (41°F).

For this example, the joint opening could be created at either the minimum curtain wall winter design temperature, $T_w$, or the summer installation temperature, $T_{is}$. If the joint opening is created at the winter installation temperature, $T_w$, the sealant will experience a dimensional change from thermal movement, $\Delta T_w$, of 74°C (133°F) for the curtain wall mullion and 85°C (152°F) for the masonry wall. If the opening is created at the summer installation temperature, $T_{is}$, the sealant will experience a dimensional change from curtain wall thermal movement, $\Delta T_s$, of 51°C (92°F) or 23°C (41°F) and for the masonry wall a dimensional change from thermal movement, $\Delta T_s$, of 51°C (92°F) or 34°C (60°F). Dimensional changes from both winter and summer installation temperature differentials will need to be evaluated to determine which condition will result in the largest joint width necessary to satisfy the performance criteria. These temperature differentials are illustrated by Fig. 6.

Both summer and winter installation conditions need to be evaluated if the joint opening installation time can not be established. However, it should be recognized that construction schedules are frequently changed and if a sealant joint is sized based on a particular installation condition that can not be met, a deficiently sized joint may be the result.

---

($T_{iw}$) Curtain wall Curtain wall Masonry

$T_w$ = -16°C (3°F) $T_{is}$ = 35°C (95°F) $T_s$ = 58°C (136°F) $T_s$ = 69°C (155°F)

|____________|____________|____________|

Winter installation:

$\Delta T_w$ = 74°C (133°F)

|------------------------------------------->| Curtain wall

$\Delta T_m$ = 85°C (152°F)

|--------------------------------------------------------------->| Masonry wall

Summer installation:

$\Delta T_s$ = 51°C (92°F) $\Delta T_s$ = 23°C (41°F)

|<-------------------+-------------------->| Curtain wall

$\Delta T_s$ = 51°C (92°F) $\Delta T_s$ = 34°C (60°F)

|<-------------------+------------------------------------------->| Masonry wall

---

FIG. 6--Expected joint opening thermal movements.

Thermal movements--Expansion joints in the masonry wall are spaced at 7.32 m (24 ft) and one-half of that wall length should contribute to extension and compression sealant joint movement. Some consultants indicate that thermal expansion of extruded face brick in a vertical direction (the direction of the coring) is perhaps 22 percent greater than in the horizontal [10, 20]. Unless independent testing provides different values to be used, it would seem prudent to include a 22 percent increase. The expected dimensional change due to the vertical, $\Delta L_{bv}$, and horizontal, $\Delta L_{bh}$, components of thermal movement for the brick wall and the curtain wall mullion vertical component, $\Delta L_{cv}$, are determined using Equation 5 as before. However, for brick wall vertical movement, it is modified for the 22 percent increase as indicated by Equation 5a.

$$\Delta L_{bv} = (L)(\Delta T_x)(\alpha_x)(1.22) \tag{5a}$$

The brick masonry dimensional changes due to the thermal movements, shown in Fig. 6 are obtained by using Equation 5 for horizontal movement and Equation 5a for vertical movement. Curtain wall vertical thermal movement is obtained using Equation 5. Values obtained from these equations are indicated in Table 6. The following is an example of a typical calculation, in this case, for the expected masonry wall horizontal dimensional change due to thermal movement for a $\Delta T_m$ of 85°C (152°F) using one-half the sealant joint spacing from Example One and Equation 5.

$$\Delta L_{bh} = 3.66(1000)(85)(0.0000065) = 2.0222 \text{ mm}$$

and

$$\Delta L_{bh} = 12(12)(152)(0.0000036) = 0.0788 \text{ in}$$

TABLE 6--Expected material dimensional changes.

| Winter Installation: | $\Delta T_w$ of 74°C (133°F); curtain wall<br>$\Delta T_m$ of 85°C (152°F); masonry | | | | |
|---|---|---|---|---|---|
| | mm | in | | mm | in |
| masonry | | | curtain wall mullion | | |
| $\Delta L_{bv}$ | 4.1117 | 0.1602 | $\Delta L_{cv}$ | 10.7433 | 0.4213 |
| $\Delta L_{bh}$ | 2.0222 | 0.0788 | | | |
| Summer Installation: | $\Delta T_s$ of 51°C (92°F); curtain wall and masonry | | | | |
| | mm | in | | mm | in |
| masonry | | | curtain wall mullion | | |
| $\Delta L_{bv}$ | 2.4670 | 0.0970 | $\Delta L_{cv}$ | 7.4042 | 0.2915 |
| $\Delta L_{bh}$ | 1.2133 | 0.0477 | | | |
| Summer Installation: | $\Delta T_w$ of 23°C (41°F); curtain wall<br>$\Delta T_m$ of 34°C (60°F); masonry | | | | |
| | mm | in | | mm | in |
| masonry | | | curtain wall mullion | | |
| $\Delta L_{bv}$ | 1.6447 | 0.0632 | $\Delta L_{cv}$ | 3.3391 | 0.1299 |
| $\Delta L_{bh}$ | 0.8089 | 0.0311 | | | |

Moisture movements--The effects of irreversible moisture movement of the brick masonry wall are again found using Equation 6 as in Example One and are for horizontal moisture movement, $\Delta L_{ih}$ = 1.0980 mm (0.0432 in); and for vertical moisture movement, $\Delta L_{iv}$ = 1.8300 mm (0.0720 in).

Movement analysis--Three combinations of vertical and horizontal dimensional change must be evaluated to determine which combination will ultimately result in the largest joint size. One set of combinations is generated by an opening that is created at the lowest or winter installation temperature, $T_{iw}$, and the other two sets by an opening that is created at the highest or summer installation temperature, $T_{is}$. The movement combinations should be evaluated both with and without the brick moisture growth effect as previously described. Fig. 7 illustrates and Table 7 lists the resulting equations for the various dimensional change combinations for this example.

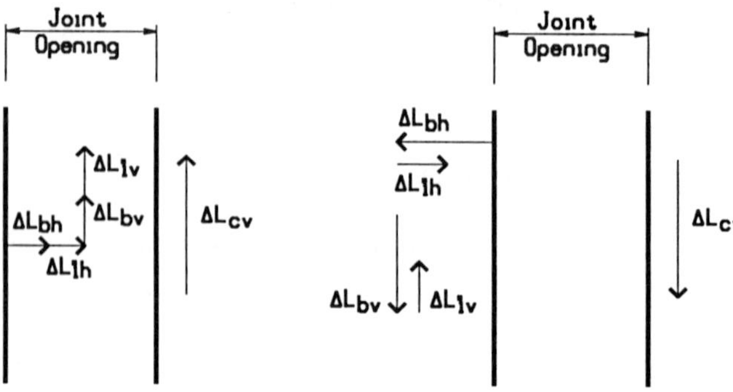


Winter Installation @ $T_{iw}$ = -16°C (3°F) and Summer Installation @ $T_{is}$ = 35°C (95°F) (warming to $T_s$)

Summer Installation @ $T_{is}$ = 35°C (95°F) (cooling to $T_w$)

FIG. 7--Movements related to installation time.

TABLE 7--Dimensional change combinations.

| | | | | | |
|---|---|---|---|---|---|
| **Winter Installation Movement Combinations** | | | | | |
| A - with moisture movement effect | | | B - without moisture movement effect | | |
| Vertical | $\Delta L_{cv} - (\Delta L_{bv} + \Delta L_{iv})$ | (9) | Vertical | $\Delta L_{cv} - \Delta L_{bv}$ | (11) |
| Horizontal | $\Delta L_{bh} + \Delta L_{ih}$ | (10) | Horizontal | $\Delta L_{bh}$ | (12) |
| **Summer Installation Movement Combinations** | | | | | |
| Cooling to $T_w$ | | | | | |
| C - with moisture movement effect | | | D - without moisture movement effect | | |
| Vertical | $\Delta L_{cv} - (\Delta L_{bv} - \Delta L_{iv})$ | (13) | Vertical | $\Delta L_{cv} - \Delta L_{bv}$ | (15) |
| Horizontal | $\Delta L_{bh} - \Delta L_{ih}$ | (14) | Horizontal | $\Delta L_{bh}$ | (16) |
| Warming to $T_s$ | | | | | |
| E - with moisture movement effect | | | F - without moisture movement effect | | |
| Vertical | $\Delta L_{cv} - (\Delta L_{bv} + \Delta L_{iv})$ | (9) | Vertical | $\Delta L_{cv} - \Delta L_{bv}$ | (11) |
| Horizontal | $\Delta L_{bh} + \Delta L_{ih}$ | (10) | Horizontal | $\Delta L_{bh}$ | (12) |

Dimensional change occurring when the joint opening is established at the winter installation temperature, $T_{iw}$, of -16°C (3°F) with a $\Delta T_m$ of 85°C (152°F) for the brick and a $\Delta T_w$ of 74°C (133°F) for the curtain wall mullions and also when the joint opening is established at the summer installation temperature, $T_{is}$, of 35°C (95°F) with a cooling $\Delta T_s$ of 51°C (92°F) for both the brick and the curtain wall mullions and a warming $\Delta T_s$ of 23°C (41°F) for the curtain wall and a $\Delta T_s$ of 34°C (60°F) for the masonry are calculated using the Equations from Table 7, the expected dimensional changes from Table 6, and the calculated moisture movements.

Dimensional changes for Table 7 movement combinations are shown in Table 8.

TABLE 8--Joint opening dimensional change calculations.

Winter Installation for: $\Delta T_m$ of 85°C (152°F) for masonry; $\Delta T_w$ of 74°C (133°F) for curtain wall

| | mm | in | | | mm | in | |
|---|---|---|---|---|---|---|---|
| A - with moisture movement effect | | | | B - without moisture movement effect | | | |
| Vertical | 4.8016 | 0.1891 | (9) | Vertical | 6.6316 | 0.2611 | (11) |
| Horizontal | 3.1202 | 0.1220 | (10) | Horizontal | 2.0222 | 0.0788 | (12) |

Summer Installation for: $\Delta T_s$ of 51°C (92°F) for masonry and curtain wall

| | | | | | | | |
|---|---|---|---|---|---|---|---|
| Cooling | | | | | | | |
| C - with moisture movement effect | | | | D - without moisture movement effect | | | |
| Vertical | 6.7672 | 0.2665 | (13) | Vertical | 4.9372 | 0.1945 | (15) |
| Horizontal | 0.1153 | 0.0045 | (14) | Horizontal | 1.2133 | 0.0477 | (16) |

Summer Installation for: $\Delta T_s$ of 34°C (60°F) for masonry; $\Delta T_s$ of 23°C (41°F) for curtain wall

| | | | | | |
|---|---|---|---|---|---|
| Warming | | | | | |
| E - with moisture movement effect | | | F - without moisture movement effect | | |
| Vertical | 0.1356 | 0.0053 | Vertical | 1.6944 | 0.0667 |
| Horizontal | 1.9069 | 0.0743 | Horizontal | 0.8089 | 0.0311 |

The following is an example of a typical calculation to arrive at a Table 8 value, in this case, for the vertical component for movement combination A from Table 7, which includes moisture movement:

$$\Delta L_{cv} - (\Delta L_{bv} + \Delta L_{iv}) \qquad (9)$$

substituting

$$10.7433 - (4.1117 + 1.8300) = 4.8016 \text{ mm}$$

and

$$0.4213 - (0.1602 + 0.0720) = 0.1891 \text{ in}$$

Since this example has more variables than Example One and thermal movement in two directions, it seems prudent to use the same sealant from Example One; however at ± 35 percent movement capacity.

TABLE 9--Joint width, $W_h$, required for horizontal dimensional change.

| | mm | in | | mm | in |
|---|---|---|---|---|---|
| **Winter Installation Movement Combinations** | | | | | |
| A - with moisture movement effect | 8.9149 | 0.3486 | B - without moisture movement effect | 5.7777 | 0.2251 |
| **Summer Installation Movement Combinations** | | | | | |
| Cooling | | | | | |
| C - with moisture movement effect | 0.3294 | 0.0129 | D - without moisture movement effect | 3.4666 | 0.1363 |
| Warming | | | | | |
| E - with moisture movement effect | 5.4483 | 0.2123 | F - without moisture movement effect | 2.3111 | 0.0889 |

Joint width required for horizontal dimensional change--The joint width required to satisfy the horizontal dimensional change component for the combinations determined above in Table 8 are found using Equation 17 as follows.

$$W_h = \Delta L_x / S \tag{17}$$

The following is an example of using Equation 17 for the horizontal component for movement combination A from Table 8.

substituting

$$W_h = 3.1202/0.35 = 8.9149 \text{ mm} \quad \text{and} \quad W_h = 0.1220/0.35 = 0.3486 \text{ in}$$

Table 9 indicates the required joint widths for movement due to horizontal dimensional change using Equation 17 and the dimensional change values from Table 8.

Joint width required for vertical dimensional change--The vertical dimensional change produces a diagonal lengthening of the sealant (Fig. 2). To not exceed the ± 35 percent movement capacity of the sealant, the diagonal length of the sealant after vertical dimensional change must not be greater than the joint width prior to the change plus 35 percent of that width. A simple trigonometric equation will provide the required joint width. If $W_v$ represents the sealant joint width prior to change, $J_l$ the vertical dimensional change, and $1.35W_v$ the diagonal sealant length after dimensional change, the equation to find the joint width required to satisfy vertical dimensional change is:

$$W_v^2 + J_l^2 = 1.35W_v^2 \qquad (18)$$

solving for $W_v$

$$W_v = \sqrt{J_l^2/0.8225}$$

The following is an example of using Equation 18 for the vertical component for movement combination A from Table 8.

substituting

$W_v = \sqrt{(4.8016)^2/0.8225} = 5.2944$ mm and $W_v = \sqrt{(0.1891)^2/0.8225} = 0.2085$ in

Table 10 indicates the required joint widths for movement due to vertical dimensional change using Equation 18 and the dimensional change values from Table 8.

TABLE 10--Joint width, $W_v$, required for vertical dimensional change.

| | | | | |
|---|---|---|---|---|
| Winter Installation Movement Combinations | | | | |
| | mm | in | mm | in |
| A - with moisture movement effect | 5.2944 | 0.2085 | | |
| B - without moisture movement effect | | | 7.3122 | 0.2879 |
| Summer Installation Movement Combinations | | | | |
| Cooling | | | | |
| C - with moisture movement effect | 7.4618 | 0.2939 | | |
| D - without moisture movement effect | | | 5.4439 | 0.2145 |
| Warming | | | | |
| E - with moisture movement effect | 0.1495 | 0.0058 | | |
| F - without moisture movement effect | | | 1.8683 | 0.0735 |

Sealant joint width--The designed width of the sealant joint is the sum of the horizontal, $W_h$, and vertical, $W_v$, joint widths as indicated in Tables 9 and 10 to which is added the construction tolerances, $C_b$ and $C_c$, using Equation 19 as follows.

$$W = W_h + W_v + C_b + C_c \qquad (19)$$

The following is an example of using Equation 19 to determine the designed joint width based on movement combination A from Tables 9 and 10.

substituting

$$W = 8.9149 + 5.2944 + 3.2 + 4.8 = 22.2093 \text{ mm}$$

and

$$W = 0.3486 + 0.2085 + 0.1250 + 0.1875 = 0.8696 \text{ in}$$

Table 11 indicates the final designed joint widths using Equation 19.

TABLE 11--Final designed joint width, W.

| | | | | | |
|---|---|---|---|---|---|
| **Winter Installation Movement Combinations** | | | | | |
| | mm | in | | mm | in |
| A - with moisture movement effect | 22.2093 | 0.8696 | B - without moisture movement effect | 21.0899 | 0.8255 |
| **Summer Installation Movement Combinations** | | | | | |
| Cooling | | | | | |
| C - with moisture movement effect | 15.7912 | 0.6193 | D - without moisture movement effect | 16.9105 | 0.6633 |
| Warming | | | | | |
| E - with moisture movement effect | 13.5978 | 0.5306 | F - without moisture movement effect | 12.1794 | 0.4749 |

The contract documents should indicate the final designed joint width based on the largest designed joint width shown in Table 11, which for this example, is produced by the winter installation movement combination that includes moisture movement. The calculated width is rounded off and the permissible construction tolerances indicated as follows.

22 mm (± 8 mm) and 7/8 in (± 5/16 in)

As is evident from this example, the influence of time of joint opening installation, expected temperature differential, and the irreversible moisture growth effect, should be evaluated for the various expected dimensional change movement combinations to determine which will produce the largest joint width design. For this example, both winter installation movement combinations with and without moisture movement produce the largest joint widths. This example indicates the importance of properly designing a joint when winter installation compressive movement is to occur.

Sealant joint depth--Refer to example one for guidelines to establish the depth of sealant required for this joint width.

Example Three: Multi-Story Building Floor Line Joint

Example description--Curtain walls on multi-story buildings are usually supported at each floor level. Horizontal sealant joints, usually called expansion joints, also occur at or near each floor level. These sealant joints must respond to many performance factors related to curtain wall design and materials, building floor loading and usage, characteristics of the building framing system, and construction tolerances among others.

This example is for a new reinforced concrete framed multi-story building in Detroit, Michigan which is clad with a light gray painted aluminum mullion and glass curtain wall system. The curtain wall has a sawtooth profile so there will be reflected solar radiation to be included in the thermal movement calculations. The curtain wall is a unitized, shop-fabricated system, with the panels being one module wide by 3.952 m (13 ft) high, which is the floor-to-floor dimension, L. The panels will be fabricated to a ±1.6 mm (1/16 in) tolerance, $C_f$, and erected to a ±3.2 mm (1/8 in) tolerance, $C_e$. For every floor level there will be a horizontal sealant joint functioning as a system expansion joint. The joint will be designed to accommodate not only thermal movement but also a differential live load deflection, $M_l$, between floors of 6.4 mm (1/4 in), lateral building sway, $M_w$, due to wind of 9.5 mm (3/8 in), and a differential floor spandrel beam deflection due to creep [1], $M_c$, of 4.8 mm (3/16 in).

TABLE 12--Performance factors - Example Three.

| Factor | Value | Factor | Value |
|---|---|---|---|
| $L$ | = 3.962 m (13 ft) | $T_{is}$ | = 38°C (100°F) |
| $C_f$ | = ± 1.6 mm (1/16 in) | $A$ | = 0.65 |
| $C_e$ | = ± 3.2 mm (1/8 in) | $\alpha_a$ | = 0.0000238 mm/mm/°C (0.0000132 in/in/°F) |
| $M_w$ | = 9.5 mm (3/8 in) | $M_c$ | = 4.8 mm (3/16 in) |
| $T_w$ | = -16°C (3°F) | $M_l$ | = 6.4 mm (1/4 in) |
| $T_a$ | = 33°C (91°F) | $S$ | = ± 50 percent |
| $T_{iw}$ | = 0°C (32°F) | | |

The above described as well as other performance factors that affect joint design are listed in Table 12. The same data sources for Example One were used.

There are performance factors other than those in Table 12 that are related to the building framing system that must be evaluated. The concrete columns will shorten due to shrinkage, creep, and elastic frame shortening. Some of these effects will occur before the curtain wall is erected and joints sealed. Theoretically, the total amount of each of these effects will not have to be compensated for in the joint design; however, determination of partial values may be difficult. It has been reported that about 50 to 75 percent of the total axial shortening, including creep, takes place before a curtain wall system is erected [21]. The author has seen a total value for these effects ranging from 3 to 6 mm (1/8 to 1/4 in) for this example's indicated joint spacing. For this example, the formwork will be constructed slightly taller in height to compensate for most of these effects so they need not be considered directly in the sealant joint design. The use of the sealant at less than its rated movement capacity should provide extra capacity to compensate for the remaining indeterminate column shortening effects. The spandrel beam, which provides support for the curtain wall, will also experience creep and shrinkage and this is expressed as a single creep value, $M_c$. The spandrel beam will be cambered to compensate for dead load deflection.

Temperature differences--The expected thermal movement is determined as follows. The surface temperature extremes are $T_w$ = -16°C (3°F) and by Equation 1, using a constant for a low heat capacity material with reflected solar radiation, $T_s$ = 80°C (176°F). Using the installation temperatures $T_{is}$ = 38°C (100°F) and $T_{iw}$ = 0°C (32°F) with Equations 3 and 4 the expected summer installation temperature differences, $\Delta T_s$, are 54°C (97°F) and 42°C (76°F) and the expected winter installation temperature differences, $\Delta T_w$, are 80°C (144°F) and 16°C (29°F). The above analysis is illustrated graphically by Fig. 8.

```
Tw = -16°C (3°F)   Tiw = 0°C (32°F)   Tis = 38°C (100°F)   Ts = 80°C (176°F)
|__________________|__________________|__________________|

ΔTw = 16°C (29°F)          ΔTw = 80°C (144°F)
|<-----------------+--------------------------------------->|
        ΔTs = 54°C (97°F)                  ΔTs = 42°C (76°F)
|<------------------------------------+-------------------->|
```

FIG. 8--Expected thermal movements.

The actual in-service temperature of the curtain wall mullions will be somewhere between the exterior winter or summer wall surface extreme and the interior ambient temperature. The actual temperature of the mullion will depend on degree of system insulation and the presence and quality of thermally improved mullions.

For information, the author, by full-size mock-up testing, has found that surface temperatures for aluminum mullions of three 4-side structural sealant glazed glass curtain walls; for summer conditions, were generally about 70 percent of the calculated summer exterior wall surface temperature in degrees fahrenheit; and for winter conditions, about 75 percent of the winter time interior air temperature in degrees fahrenheit. For this example, for an occupied and conditioned building, a 176°F outside summer surface temperature would result in a mullion temperature for calculation of 132°F and for winter conditions a 75°F internal air temperature would result in a mullion temperature for calculation of 56°F.

These values could be used in lieu of the above temperature extremes to calculate the joint width. However, use of the expected extreme surface temperatures will be conservative and provides some insurance if the building is unheated or uncooled for a period of time (i.e. power failure), and if the building thermostats may be set back to a minimum value over weekends or other extended periods.

Thermal movements--The expected curtain wall mullion thermal movements, $\Delta L_{cv}$, that will occur for the above temperature differences are found using L = 3.962 mm (13 ft), $\alpha_a$, and Equation 5 and for a $\Delta T$ of 80°C (144°F) are 7.5436 mm (0.2965 in); for a $\Delta T$ of 16°C (29°F) are 1.5087 mm (0.0597 in); for a $\Delta T$ of 54°C (97°F) are 5.0920 mm (0.1997 in); and for a $\Delta T$ of 42°C (76°F) are 3.9604 mm (0.1565 in).

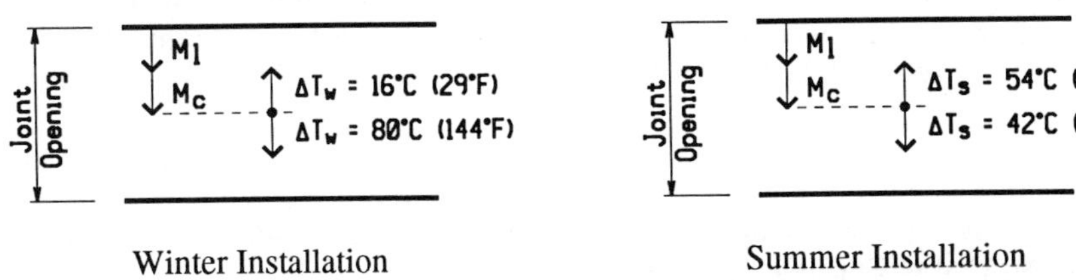


FIG. 9--Movements related to installation time.

Movement analysis--For winter installation conditions, as illustrated by Fig. 9, the joint is designed for the effects of live load, creep, and thermal movement, which is additive, reducing the joint opening, as the building warms to the expected summer maximum and is subtractive, tending to increase the joint opening, as the building cools to the expected winter minimum.

For summer installation conditions, as illustrated by Fig. 9, the joint is designed for the effects of live load, creep, and thermal movement, which is additive, reducing the joint opening, as the building warms to the expected summer maximum and is subtractive, tending to increase the joint opening, as the building cools to the expected winter minimum. After the joint width is established for the above movement conditions it will be checked to see if it will be able to accommodate the effects of building lateral sway due to wind. The following illustrates the required calculations.

Some spandrel beam creep, $M_c$, estimated at about 40 percent of the value provided in Table 12, will occur before erection of the curtain wall system. For this application the creep value for calculation will be: 4.8(0.60) = 2.8800 mm and 0.1875(0.60) = 0.1125 in.

As Fig. 9 illustrates, the in-plane movements of $M_l$, $M_c$, and $\Delta L_{cv}$ that the joint will experience are additive and in some cases subtractive. Table 13 lists these values and indicates the net joint movement, $\Delta L$, for the expected $\Delta T$'s .

TABLE 13--Joint movement effects.

| for $\Delta T$ of: | 80°C (144°F) mm in | 16°C (29°F) mm in | 54°C (97°F) mm in | 42°C (76°F) mm in |
|---|---|---|---|---|
| $M_l$ | 6.4000 (0.2500) | 6.4000 (0.2500) | 6.4000 (0.2500) | 6.4000 (0.2500) |
| $M_c$ | 2.8800 (0.1125) | 2.8800 (0.1125) | 2.8800 (0.1125) | 2.8800 (0.1125) |
| $\Delta L_{cv}$ | 7.5436 (0.2965) | -1.5087 (-0.0597) | -5.0920 (-0.1997) | 3.9604 (0.1565) |
| net $\Delta L$ | 16.8236 (0.6590) | 7.7713 (0.3028) | 4.1880 (0.1628) | 13.2404 (0.5190) |

The same sealant for the previous examples will be used at ± 40 percent movement capacity. The joint width required to satisfy $\Delta L$, the net joint movement effect provided in Table 13, is found using Equation 20.

$$W_m = \Delta L/S \tag{20}$$

The following is an example of using Equation 20 to determine the net movement for a $\Delta T$ of 80°C (144°F) using Table 13.

substituting

$$W_m = 16.8236/0.40 = 42.0590 \text{ mm} \quad \text{and} \quad W_m = 0.6590/0.40 = 1.6475 \text{ in}$$

Table 14 indicates the required joint widths for movement due to in-plane dimensional change using Equation 20 and the net $\Delta L$ values from Table 13.

TABLE 14--Required joint widths, $W_m$, for in-plane movement effects.

| for $\Delta T$ of: | 80°C (144°F) mm | in | 16°C (29°F) mm | in | 54°C (97°F) mm | in | 42°C (76°F) mm | in |
|---|---|---|---|---|---|---|---|---|
| $W_m$ is: | 42.0590 | 1.6475 | 19.4283 | 0.7570 | 10.4700 | 0.4070 | 33.1010 | 1.2975 |

Wind drift, $M_w$, the per story lateral drift or sway of the building, can be compensated for with a trigonometric formula similar to that used in Example Two. The joint after movement can not be extended greater than 40 percent of its width prior to movement. By using $M_w$ and Equation 21 the joint width, $W_w$, required for wind drift lateral movement can be determined.

$$W_w^2 + M_w^2 = 1.4W_w^2 \qquad (21)$$

solving for $W_w$

$$W_w = \sqrt{M_w^2/0.96}$$

substituting

$$W_w = \sqrt{(9.5)^2/0.96} = 9.6959 \text{ mm} \quad \text{and} \quad W_w = \sqrt{(0.3750)^2/0.96} = 0.3827 \text{ in}$$

Based on the joint opening width, $W_w$, necessary for wind drift the maximum allowable sealant extension is 9.6959(0.40) = 3.8784 mm and 0.3827(0.40) = 0.1531 in.

Review of the net values for $\Delta L$ ( Table 13) indicates that the joint opening is in compression for the referenced $\Delta T$'s. If the $\Delta L$ values in compression are in excess of the movement or sealant extension required for wind drift then the size of the joint opening for movement will be the values found by Equation 20 and indicated in Table 14. The joint by being in a closed state has a latent capacity to accommodate the anticipated lateral movement due to wind drift.

The final designed width, W, of the sealant joint is determined according to Equation 22 using the largest value for $W_m$ from Table 14.

$$W = W_m + C_f + C_e \qquad (22)$$

substituting

$$W = 42.0590 + 1.6 + 3.2 = 46.8590 \text{ mm} \quad \text{and} \quad W = 1.6475 + 0.0625 + 0.1250 = 1.8350 \text{ in}$$

and rounding off

$$W = 47 \text{ mm (1-13/16 in)}$$

The contract documents should indicate the final designed joint width with the permissible fabrication and erection construction tolerances as follows.

47 mm (± 5 mm) and 1-13/16 in (± 3/16 in)

Refer to Example One for guidelines to establish the depth of sealant required for this joint width. If depth is not properly established the ability of the sealant to accommodate movement can be seriously compromised.

Example Four: Fillet Joint Between Different Materials or Systems

Butt type sealant joints are traditionally used for joints that experience movement. Fillet type joints, illustrated by Fig. 10, have been used; however, little theoretical information was available for performance during movement, to assist the joint designer in determining the size of the sealant fillet, and configuration requirements for the sealant backing. Recently, research information was published describing fillet sealant joint performance [22, 23]. The following movement analysis is based on that work.

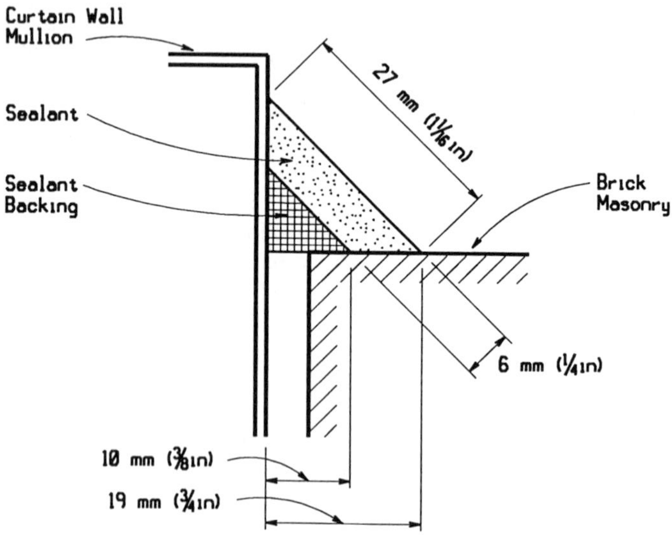


FIG. 10--Fillet type sealant joint profile.

Often, when failed sealant joints are encountered on existing buildings, joint designers are confronted with only one joint profile, a fillet, as the most economical, or due to construction constraints, the only available sealant profile to use for repair of a joint that experiences movement. Typically, these joints can be found at transition areas, such as, the perimeter of window frames and metal framed glazing systems to other systems such as masonry walls.

The criteria from Example Two will be used for this example. The Example Two building was constructed without design tolerances being considered for the sealant joint between the lobby curtain wall and adjacent masonry wall. Also, the joint was not properly designed for movement and soon after occupancy the joint failed. Economic considerations have resulted in the decision to reseal the joint using a fillet type remedial sealant joint. Relevant performance factors are listed in Table 15.

TABLE 15--Performance factors - Example Four.

| Factor | Value | Factor | Value |
|---|---|---|---|
| $L$ = | 6.1 m (20 ft) | $\alpha_a$ = | 0.0000238 mm/mm/°C (0.0000132 in/in/°F) |
| $T_i$ = | 20°C (68°F) | | |
| $T_s$ = | 58°C (136°F) | $S$ = | ± 50 percent |

Thermal movement of the components will be determined based on an occupied and conditioned building as follows.

For this example, the curtain wall was constructed using a 4-side structural sealant glazed glass and metal framing system. As described earlier, the author, by mock-up testing, has found that the surface temperatures for the aluminum mullions of three 4-side structural sealant glazed glass curtain walls for summer conditions was generally about 70 percent of the calculated summer surface temperature and for winter conditions about 75 percent of the winter time interior air temperature.

Also, for this example, the effect of thermal movement of the masonry wall will not be considered since the masonry will experience a time lag for absorbing heat and increasing in dimension due to its thermal mass. Irreversible moisture movement will also not be considered since the wall has been erected for some time and most of the irreversible growth has occurred. Both thermal and irreversible movement of the masonry wall would have countering effects to curtain wall movement, so the calculated results will be slightly conservative.

From Example Two, the summer surface temperature of the curtain wall, $T_s$, is 58°C (136°F) and, for this example, the winter interior air temperature, $T_i$, for the occupied and conditioned building is 20°C (68°F). Using the above mock-up testing criteria, both the $T_s$ and $T_i$ fahrenheit temperatures should be multiplied by 70 and 75 percent respectively to yield values to determine the expected occupied and conditioned building temperature difference for sealant joint design. These values and the equivalent centigrade temperatures are for, $T_s$, 35°C (95°F) and for, $T_w$, 11°C (51°F). The expected curtain wall temperature difference, $\Delta T_m$, is 24°C (44°F) according to Equation 2.

Thermal movement of the curtain wall based on the $\Delta T_m$ of 24°C (44°F) is found using Equation 5 as follows.

$$\Delta L_{cv} = 6.1(1000)(0.0000238)(24) = 3.4843 \text{ mm}$$

and

$$\Delta L_{cv} = 20(12)(0.0000132)(44) = 0.1394 \text{ in}$$

According to reference [22], the length of the contact side of the triangular shaped sealant backing, B, is determined by dividing the thermal movement by the movement capacity of the sealant as illustrated by Equation 23. For this example, the joint design is for as-built conditions where optimum surface adhesion may be difficult to achieve due to the difficulty of adequately cleaning the original sealant from the remedial sealant work adhesion surfaces. Due to these conditions, the ± 50 percent movement capacity sealant will be used at a ± 35 percent rating.

$$B = \Delta L_x / S \qquad (23)$$

substituting $\Delta L_{cv}$ of 3.4843 mm (0.1394 in),

$$B = 3.4843/0.35 = 9.9551 \text{ mm} \quad \text{and} \quad B = 0.1394/0.35 = 0.3983 \text{ in}$$

and rounding off,

$$B = 10 \text{ mm } (3/8 \text{ in})$$

As illustrated by Fig. 10, the sealant designed thickness should be 6 mm (1/4 in) to ensure that, during application, no less than a 3 mm (1/8 in) thickness is achieved. Based on the determined size of the sealant backing and the minimum sealant thickness the joint profile is 21 mm (13/16 in) along the contact legs and 29 mm (1-1/8 in) across the exposed face. It is good practice to limit the length of the contact face to a maximum of 25 mm (1 in) to facilitate sealant installation. Additional information relative to fillet joint design is available from references [22 ] and [23].

## Example Five: Bridge Joint in a Precast Concrete Panel System

As in Example Four, the joint designer is often faced with existing joints that have failed and need to be resealed. For this example, a precast concrete panel system on a building in Detroit, Michigan has deficiently sized butt type sealant joints that occur between the vertical edges of the panels. The joints have failed due to inadequate consideration of fabrication and erection tolerances. It is not economically viable to enlarge the joint opening so a bridge type sealant joint (colloquially known as a band-aid joint) will be applied to the panel faces to increase the size of the sealant joint and therefore its movement capability.

Butt type sealant joints traditionally have been used for joints that are going to experience movement. Other joint shapes, such as a bridge illustrated by Fig. 11, have been used; however, little information was available related to performance during movement to assist the joint designer in determining the size of the sealant bridge and configuration requirements for bond-breaker tape or sealant backing. Recently, research information was published describing bridge type sealant joint performance [24]. The following movement analysis is based on that work.

The precast concrete panels are 1.524 m (5 ft) high and 9.144 m (30 ft) long and have a light color finish. It has been determined that the panels are attached at both the bottom and top of the panels with a fixed connection at the top of the panel at a building column. The building insulation system is located inward of the panel connections to the building columns and therefore the panels and connections will experience close to the expected temperature extremes. The above, as well as other performance factors that will affect joint design, are listed in Table 16.

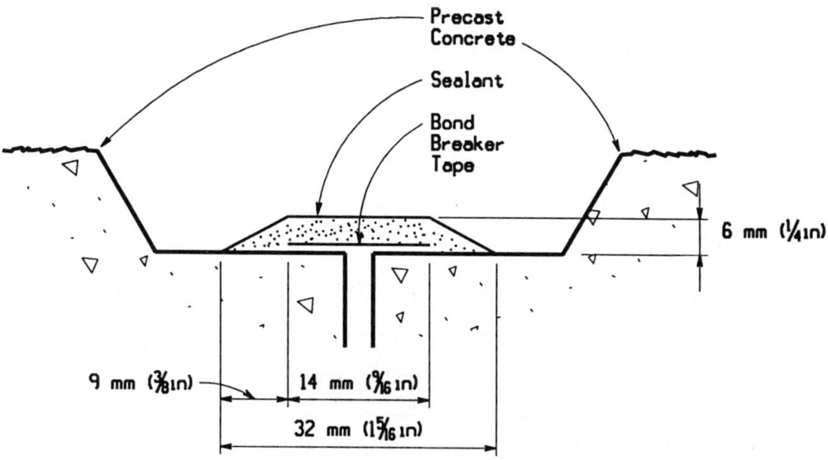


Fig. 11--Bridge type sealant joint profile.

TABLE 16--Performance factors - Example Five.

| Factor | Value | Factor | Value |
|---|---|---|---|
| L | = 9.144 m (30 ft) | $T_{is}$ | = 35°C (95°F) |
| $T_w$ | = -16°C (3°F) | A | = 0.65 |
| $T_a$ | = 33°C (91°F) | $\alpha_c$ | = 0.0000108 mm/mm/°C (0.000006 in/in/°F) |
| $T_{iw}$ | = 4°C (40°F) | | |

The expected precast panel dimensional change due to thermal movement is determined as follows. The hottest summer wall surface temperature, $T_s$, is found by Equation 1, using a constant for a high heat capacity material.

$$T_s = T_a + 42S = 33 + 42(0.65) = 60°C \quad \text{and} \quad T_s = T_a + 75S = 91 + 75(0.65) = 140°F$$

The coldest winter wall surface temperature, $T_w$, is -16°C (3°F). For this example, values for $T_{is}$ and $T_{iw}$ for the sealant installation are known, and the expected temperature differences, $\Delta T_s$ and $\Delta T_w$ are found using Equations 3 and 4.

substituting, summer installation temperature difference using Equation 3

$$\Delta T_s = 35 - (-16) = 51°C \quad \text{and} \quad \Delta T_s = 95 - 3 = 92°F$$

and winter installation temperature difference using Equation 4

$$\Delta T_w = 60 - 4 = 56°C \quad \text{and} \quad \Delta T_w = 140 - 40 = 100°F$$

The above analysis is expressed graphically by Fig. 12.

$T_w$ = -16°C (3°F) $T_{iw}$ = 4°C (40°F) $T_{is}$ = 35°C (95°F) $T_s$ = 60°C (140°F)

|_______________|_______________|_______________|

$\Delta T_w$ = 56°C (100°F)

|------------------------------------------>|

$\Delta T_s$ = 51°C (92°F)

|<------------------------------------------|

FIG. 12--Expected thermal movements.

The change in dimension for precast panel thermal movement is greatest for sealant winter installation and is determined using Equation 5.

substituting, using the winter installation temperature difference $\Delta T_w$ of 56°C (100°F)

$$\Delta L_w = (9.144)(1000)(56)(0.0000108) = 5.5303 \text{ mm}$$

and

$$\Delta L_w = (30)(12)(100)(0.000006) = 0.2160 \text{ in}$$

According to reference [24], the length of the bond-breaker tape or sealant backing, B, is determined by dividing the expected thermal movement by the movement capacity of the sealant as illustrated by Equation 23. For this example, the adhesion surfaces for the remedial sealant work can be readily cleaned of old sealant and other materials permitting the 50 percent movement capacity sealant to be used at a 40 percent rating.

substituting $\Delta L_w$ of 5.5303 mm (0.2160 in),

B = 5.5303/0.40 = 13.8258 mm and B = 0.2160/0.40 = 0.5400 in

and rounding off,

B = 14 mm (9/16 in)

As illustrated by Fig. 11, the sealant designed thickness over the bond-breaker tape or sealant backing should be 6 mm (1/4 in) to ensure that, during application, no less than a 3 mm (1/8 in) thickness is achieved. Based on the determined size of the bond-breaker tape and the minimum sealant adhesion contact widths the joint is 32 mm (1-5/16 in) wide across its exposed face. Additional information relative to bridge joint design and configuration is available from reference [24].

## CONCLUSION

The design professional, as the examples illustrate, should evaluate each sealant joint from both qualitative and quantitative positions for sealant joint performance factors, as well as the type(s) of movement to be accommodated. Also, it is advisable to not use the sealant at its rated capacity but at a lesser value. Doing so provides some insurance since there is a degree of uncertainty in establishing the magnitude of many of the performance factors and tolerances. The design professional, using the simple mathematical approximations illustrated in this paper, can design sealant joints for movement and other effects that will preclude many of the aforementioned failures which often lead to shortened building material or system durability.

## ACKNOWLEDGEMENT

The author would like to recognize the efforts of Mr. Erich Hiedemann, AIA for the preparation of the computer generated figures for this paper.

## REFERENCES

[1] "C717-94, Terminology of Building Seals and Sealants", Annual Book of ASTM Standards, Volume 04.05, ASTM, Philadelphia, 1994.

[2] Klosowski, J.M., Sealants in Construction, Marcel Dekker, Inc., New York, 1989.

[3] O'Connor, T.F., "Specifying Weather Sealants", Progressive Architecture, April, 1993, PP 39-42.

[4] "C1193-91", Standard Guide for Use of Joint Sealants", Annual Book of ASTM Standards, Volume 04.07, ASTM, Philadelphia, 1994.

[5] O'Connor, T.F., "Design of Sealant Joints", ASTM STP1069 Building Sealants: Materials, Properties, and Performance, August, 1990, American Society for Testing and Materials.

[6] O'Connor, T.F., "Exterior Wall Sealant Joint Design", Progressive Architecture, November, 1991, PP 121-125.

[7] "Table 1, Climatic Conditions for the United States", Chapter 24, American Society of Heating, Refrigerating and Air Conditioning Engineers, Inc. (ASHRAE) Handbook of Fundamentals, ASHRAE, Atlanta, 1993.

[8] Handegord G.O., and Karpati, K.K., "Joint Movement and Sealant Selection", Canadian Building Digest 155, National Research Council of Canada, Ottawa, 1973.

[9] Latta, J.K.," Dimensional Changes Due to Temperature", Record of the DBR Building Science Seminar on Cracks, Movements and Joints in Buildings, NRCC 15477, National Research Council of Canada, Ottawa, 1972.

[10] Grimm, C.T., "Thermal Strain in Brick Masonry", Proceedings of the Second North American Masonry Conference, The Masonry Society, 1982.

[11] Window 4.0, A Fenestration Product Thermal Analysis Program, Windows and Daylighting Group, Energy and Environment Division, Lawrence Berkeley Laboratory, Berkeley, California, 1988-1992.

[12] "Estimation of Thermal and Moisture Movements and Stresses: Part 1", Building Research Establishment Digest 227, Building Research Station, Garston, Watford UK, 1979.

[13] "Estimation of Thermal and Moisture Movements and Stresses: Part 2", Building Research Establishment Digest 228, Building Research Station, Garston, Watford UK, 1979.

[14] Building Code Requirements for Masonry Structures (ACI 530-88/ASCE 5-88) and Specifications for Masonry Structures (ACI 530.1-88/ASCE 6-88), American Concrete Institute and American Society of Civil Engineers, Detroit and New York, 1988.

[15] Building Movements and Joints, Portland Cement Association, Skokie, 1982.

[16] Rainger, P., Movement Control in the Fabric of Buildings, Nichols Publishing Company, New York, 1983.

[17] "C216-91c, Standard Specification for Facing Brick (Solid Masonry Units Made From Clay or Shale)", Annual Book of ASTM Standards, Volume 04.05, ASTM, Philadelphia, 1991.

[18] Manual for Quality Control for Plants and Production of Architectural Precast Concrete Products, MNL-177-77, Prestressed Concrete Institute, Chicago, 1977.

[19] Shisler, F.W. and Klosowski, J.M., "Sealant Stresses in Tension and Shear", ASTM STP 1069 Building Sealants: Materials, Properties, and Performance, August, 1990, American Society for Testing and Materials.

[20] Grimm, C.T., "Designing Brick Masonry Walls to Avoid Structural Problems", Architectural Record, October, 1977, PP 125-128.

[21] Korista, D.S., "Exterior Facade System - Building Structure Systems - Load-Deformation Behavioral Inter-Relationships", Exterior Claddings on High-Rise Buildings, Proceedings of the Chicago Committee on High Rise Buildings, 1989 Fall Symposium, Chicago, IL. 1990.

[22] Myers, J.C., "Behavior of Fillet Sealant Joints", ASTM STP 1069 Building Sealants: Materials, Properties, and Performance, August, 1990, American Society for Testing and Materials.

[23] Myers, J.C., Behavior of Fillet Sealant Joints, Master's Thesis, Department of Materials Science and Engineering, Massachusetts Institute of Technology, Cambridge, MA, June 1989.

[24] Myers, J.C., "Performance of Band-Aid Sealant Joints", ASTM STP1243 Science and Technology of Building Seals, Sealants, Glazing, and Waterproofing - Fourth Volume, David H. Nicastro Ed., ASTM, Philadelphia 1994.

James C. Myers[1]

# A COMPARISON OF DIFFERENT SEALANT BUTT JOINT SHAPES

---

**REFERENCE:** Myers, J. C., "**A Comparison of Different Sealant Butt Joint Shapes,**" Science and Technology of Building Seals, Sealants, Glazing, and Waterproofing: Fifth Volume, ASTM STP 1271, Michael A. Lacasse, Ed., American Society for Testing and Materials, 1996.

**ABSTRACT:** This paper reviews the evolution of the butt sealant joint from the deep rectangular joints used in early highway seals to the hourglass shapes used presently in building seals. The paper compares the relative performance of several types of butt sealant joints, including square, rectangular, and hourglass. The performance measures include tensile and cyclic laboratory tests and finite element analyses. The paper compares the performance of hourglass joints in recent tests to the performance of butt joints with other shapes presented in previous publications. These data can assist ASTM Committee C-24 in evaluating the effect of joint shape on the performance of specimens in tensile adhesion or cyclic tests.

**KEYWORDS:** butt joint, hourglass joint, joint shape, movement capability, sealant joint design, sealant joint stress.

---

For the past several years, various subcommittees of ASTM C-24 have been debating the effect of butt joint shape on performance capability. The question arose as latex sealants attempted to qualify in the ASTM Standard Specification for Elastomeric Joint Sealants (C 920). Latex sealants tend to shrink during cure and develop concave surfaces, unlike most chemically curing sealants. Some committee members argued that this provided an unfair advantage to the latex sealants during testing under the ASTM Test Method for Adhesion and Cohesion of Elastomeric Joint Sealants Under Cyclic Movement (Hockman Cycle) (C 719). They felt the change in geometry would improve performance relative to the standard square joint shape specified in most ASTM building sealant test methods.

The issue of joint shape arose again during discussions of the appropriate joint shape to be using during the test for movement during cure as the task group developed

---

[1]Associate, Simpson Gumpertz & Heger Inc., 297 Broadway, Arlington, MA 02174.

the method. Proponents of testing the hourglass shape, as opposed to the square butt joint, reasoned that they should be testing the same joint shape as was typically being installed in building joints. Other members wanted to carry that reasoning further and change the joint shape in the Hockman cycle (ASTM C 719) from a square to a hourglass butt joint. Some members questioned the appropriateness of a specimen that was only 51-mm (2-in.) long.

To help answer some of these questions, this paper reviews the effect of butt joint shape on sealant performance. The paper starts by reviewing the historical development of the hourglass shape and 51-mm (2-in.) specimen length that arose from research in the 1960's and early-1970's. It supplements these early findings with my more recent research conducted during an evaluation of fillet sealant joint behavior. The paper then examines the actual performance of different butt joint shapes during tensile and cyclic tests, as well as the internal stress distribution based on finite element analyses. I then present my conclusions and recommendations relative to ASTM Committee C-24's selection of joint shapes for various test methods.

## EARLY SEALANT JOINT RESEARCH

Egons Tons at MIT, John Cook at Rensselaer Polytechnic Institute, and E. Catsiff et al. at Thiokol Chemical Corporation contributed to the promotion of the hourglass butt joint as the ideal joint shape. Some of them also evaluated the effects of joint specimen length.

### Egons Tons' Research

In 1959 and 1963, Tons [1] [2] published research he had conducted on the relationship between joint width, sealed depth, and joint expansion for highway joint seals. He concluded the following:

- Upon extension, the surface of rectangular joints follows a parabolic curve.
- The maximum strain in the joint can be calculated as the ratio of the length of the parabolic face to the unstrained joint width.
- For a given depth with equal percentage extensions, joints that are wider have smaller maximum strains.
- Shallower joints have less strain than deeper joints of the same width for a given extension.

This research was a cornerstone in the development of the hourglass shape as the preferred butt joint configuration. Previously, designers had operated under the unconfirmed assumption that deeper joints performed better than shallower joints. The

Tons model is most applicable to ideal rubber materials (i.e., incompressible or poisson's ratio approximately equal to 0.5). Most of the sealants at the time of his research were not highly elastic, but were plastic or visco-elastic and subject to stress relaxation. His research is especially applicable to today's highly elastic sealants. Nevertheless, the theory and its conclusions are valuable. Using simple mathematical relationships he reached the same conclusions as other later researchers who used sophisticated finite element calculations and computers, e.g., Thiokol's research noted below. This work began the trend toward shallower joint seals.

John Cook's Research

Cook considered Ton's research and examined two related aspects for work he undertook in 1965 [3] [4]. Cook found that polysulfide sealants could be modeled more accurately as viscoelastic materials than as perfectly elastic materials. With stress relaxing materials, he found that the problem "becomes one of shape and recovery rather than stress" [3]. Viscoelastic behavior compounds the analysis of the joint behavior and tends to diminish the importance of determining the stresses in the sealant. Today, many of the high performance sealants are closer to perfectly elastic materials. The stresses that develop as these elastic joints are strained are important, since these sealants relax less with time.

Cook also examined the effect of specimen length on stresses in the sealant. This issue has arisen frequently during recent ASTM C-24 discussions regarding appropriate joint shape and size. Cook evaluated 51-mm (2-in.) and 152-mm (6-in.) long joints. He observed that end effects disturbed the full length of the shorter specimens, while they did not disturb the center half of the longer specimens. The modulus and ultimate (nominal) tensile stress in both length joints were nearly the same.

Thiokol's Research

In 1970 and 1971, Catsiff et al. examined rectangular sealant joints using finite element analysis [5] [6] [7]. They accounted for the nonlinear viscoelastic character of the sealant materials and for the finite strains involved. This was the first sophisticated examination of the stress distribution within sealant joints. They found the following:

- The nominal tensile stresses predicted by the finite element analyses agree reasonably well with laboratory test results for 152-mm (6-in.) specimens and overestimate the laboratory test results by about one-third for 51-mm (2-in.) long specimens.

- The strain distribution along the sealant surface is nonuniform, creating a non-parabolic surface with a flat center area and sharp curves at the edges of the joint. These computer predictions agreed well with their laboratory measurements.

- Maximum principal stresses are highest at the corners of the joints.

- The analyses predict that joints with convex surfaces (bulging) can have three times greater peak stresses at 25% extension than joints with flat surfaces.

This last finding begins to indicate the importance of joint shape. Unfortunately, Catsiff et al. did not continue their work to compare the stresses in hourglass joints to rectangular butt joints. The industry, however, headed in the right direction toward shallower joints and eventually settled on the hourglass shape as the optimum joint without further detailed research.

## RECENT RESEARCH BY THE AUTHOR

I studied the effect of joint shape on sealant performance and stress distribution within the joint as part of research into fillet sealant joint behavior [8] [9]. That research included various butt joint shapes in the computer analyses, but lacked an hourglass shape in the laboratory testing since the focus at the time was on fillet joints. Subsequently, I included a hourglass joint in similar laboratory testing and report those results here. Table 1 describes each of the butt joint shapes used in this research.

TABLE 1 -- Joint shapes.

| Joint Shape | Joint Width | Centerline Depth | Bondline Depth |
|---|---|---|---|
| **Deep rectangular** | 12.7 mm (0.5 in.) | 25.4 mm (1.0 in.) | 25.4 mm (1.0 in.) |
| **Square** | 12.7 mm (0.5 in.) | 12.7 mm (0.5 in.) | 12.7 mm (0.5 in.) |
| **Hourglass** | 12.7 mm (0.5 in.) | 6.4 mm (0.25 in.) | 12.7 mm (0.5 in.) |
| **Shallow rectangular** | 12.7 mm (0.5 in.) | 6.4 mm (0.25 in.) | 6.4 mm (0.25 in.) |
| **Shallow hourglass** | 12.7 mm (0.5 in.) | 3.2 mm (0.125 in.) | 6.4 mm (0.25 in.) |

### Computer Analyses

The computer analyses I reported in Reference [8] provided a non-dimensional ranking of the peak maximum principal stress in various butt sealant joints. Table 2 summarizes the non-dimensional peak stress ranking. The deep rectangular joint was arbitrarily assigned a value of 10.0 to create relative comparison rankings with the other joint shapes. The peak stress always occurred at the corners of the joint (as viewed in cross section), except for the hourglass shape where the peak stress was on the joint surface along the centerline of the joint.

TABLE 2 -- Peak stress rankings.

| **Joint Shape** | **Peak Stress Ranking** |
|---|---|
| **Deep rectangular** | 10.0 |
| **Square** | 6.3 |
| **Hourglass** | 2.6 |
| **Shallow rectangular** | 4.2 |
| **Shallow hourglass** | 2.6 |

While these analyses are not necessarily suitable for predicting the exact internal stresses in the joints, they are a good indication of the relative peak stress in the joints. These analyses show the following:

- Both the depth and the surface profile (flat vs. concave) have a sizable impact on the peak stress in a joint.
- For a square joint, the impact on peak stress of creating concave faces (hourglass) is greater than the impact of reducing the joint depth (shallow rectangular).
- These rankings tend to indicate that substantially lower peak stresses exist in hourglass or shallow joints, compared to the square joint.
- The hourglass joint has the lowest stresses along the bondline of these five joint shapes, due to the shift in location of the peak stress away from the bondline.

## Laboratory Tests

I tested various sealants using the square sealant joint in tensile and cyclic tests and the shallow rectangular sealant joint in cyclic tests, as described in References [8] and [9]. In 1993 I tested one of the same medium-modulus silicone sealants as used in the earlier research using a hourglass joint in similar tensile and cyclic tests. Table 3 summarizes the results of the tensile tests.

Somewhat surprisingly, the hourglass joint extends only about one-third as far as the square joint. The hourglass joint splits cohesively in the center of the joint, as the computer analyses would suggest. The square joints failed along the bondline, although they generally left a thin layer of sealant on the substrate. The increase in specimen length does not alter the strains or unit loads appreciably.

TABLE 3 -- Tensile test results.

| Joint Shape | Average Ultimate Nominal Strain, % | Average Ultimate Load, N |
|---|---|---|
| **Square** - 51 mm length | 289 | 404 |
| **Square** - 152 mm length | 284 | 1284 |
| **Hourglass** | 89 | 325 |

Table 4 summarizes the results of the cyclic tests. The cyclic testing did not produce definitive results because of the limited number of strains tested. The tests show that, if there is any improvement in performance attributable to the hourglass or shallow rectangular joint shapes over the square joint, it is not sufficient to increase the ASTM C 719 rating from 25% to 50% for the particular sealant tested here.

TABLE 4 -- Cyclic test results.

| Joint Shape | Tested Strains, % | Max. Passing Strain, % |
|---|---|---|
| **Square** | 12.5, 25, 37.5, 50 | 25 |
| **Hourglass** | 50, 75 | below 50 |
| **Shallow rectangular** | 50, 75 | below 50 |

## RECENT RESEARCH BY ASTM COMMITTEE C-24

As part of the development of research data on latex sealant performance over the past few years, Subcommittee C-24.16 tested various latex sealants in the Hockman Cycle (ASTM C 719) [10]. Part of the work involved testing the same sealant in two different configurations. In one test, the joint substrates were blocked apart during cure, causing the sealant to adopt a hourglass shape as it cures. In the companion test, the joint substrates were not blocked apart during cure, allowing the samples to maintain a roughly square configuration. The testing found that, for a given latex sealant, both joint shapes passed the 12.5% rating of ASTM C 719.

## DISCUSSION

The following discusses the above-summarized results with respect to selection of joint specimen length and joint specimen shape for ASTM C-24 test standards.

Joint Specimen Length

Both my tensile tests and Cook's tests found little affect on the nominal behavior of the sealant joint due to changes in test specimen length. Despite nearly the entire length of the shorter specimens being affected by the end effects, the shorter joints reach ultimate loads and strains similar to that for the longer specimens. Little, if any, testing has been done on the effect of specimen length during cyclic tests. Most likely, the testing standards and laboratories have tended toward the shortest possible specimen because of the following:

- increased loads with larger specimens makes manual cycling difficult and creates the need for stronger equipment
- increased loads also need stronger substrates, and
- the lack of standard testing equipment capable of accommodating longer specimens.

The final selection of an appropriate specimen length depends primarily on the nature of the test. If the test is designed solely to elicit nominal ultimate stresses and strains, then the past research indicates that 51 mm specimens are adequate. The appropriate length for cyclic tests, such as the Hockman Cycle, is not a well-researched subject. However, practical limitations prevent using longer specimens. The current body of test results on shorter specimens appears to be producing adequate field performance, but the short specimens could be producing conservative results.

Joint Specimen Shape

As researchers in the 1960's discovered, shallower joints perform better than deeper joints for a given joint width. The industry further refined this concept with the evolution of the hourglass butt joint as the optimum joint shape. The more difficult question to answer with today's sealants is how much improvement in performance can a particular shape provide. Computer analyses suggest that the reduction in peak stress is substantial with changes from square to hourglass or shallow rectangular joints.

Tensile and cyclic test results do not show the improvement in performance that the computer results suggest. These initial test results need to be treated with some caution. The tests did not involve a wide variety of sealants either within or among chemical classes. Sealant joint performance is a function of many variables, including material cohesive strength, adhesive strength, tear resistance, modulus, and other properties. The data thus far suggest that the degree of performance improvement attributable to hourglass or shallow rectangular joint shapes over square joints is not enough to boost the ASTM C 719 rating from 25% to 50% for at least one particular sealant. Similar results may not hold for other sealants, particularly if the formulated balance between adhesive and cohesive strengths is not similar to that of the medium-modulus silicone sealant included in the tests.

Hourglass shape joints appear to be most beneficial with sealants that have low adhesive strengths. The hourglass configuration not only lowers the peak stress, compared to a square joint, it shifts the peak stress into the body of the sealant material away from the bondline. The exact shape of the joint is a critical factor in causing the location of peak stress to shift. Shallow rectangular and shallow hourglass joints do not undergo this transition, and the peak stress in them remains along the bondline. Formulating a sealant material that relied on this transition to perform adequately would be risky, due to the ease with which a slight alteration by an applicator could eliminate the critical shape.

With regard to laboratory testing and reproducibility of test results, the square joint is superior to the hourglass joint. The exact depth and profile of the hourglass create additional variables that do not exist with square joint shapes. A shallow rectangular shape would tend to avoid these variables and create a joint more representative of optimum shape. However, the shallow rectangular joint is likely to exhibit more varied results, because slight differences in joint thickness should have a larger impact on performance than they would with square joints.

Many of the ASTM sealant test methods are not highly precise tests, as the results can sometimes vary widely. Given this uncertainty, the joint shapes being tested should be those that produce the most conservative results. The current approach of using square specimens provides some inherent cushion or factor of safety in the use of the test results, if the applicator produces a shallower or hourglass-shaped joint. If the optimum joint shape is tested and the results used without any factor of safety, then slight deviations in the installation could produce unfavorable results. The most rational approach would be to test specimens with one shape and then apply an appropriate factor of safety in the sealant joint design, based on the circumstances surrounding the application and the materials involved. Development of such a rational design approach could be the subject of many more papers.

In some test methods, it would be appropriate to use the same joint shape in the laboratory testing as will be used in the field. For example, test methods are being developed to evaluate the effects of movement during cure. The joint shape may influence the rate of cure of some sealants. The shape may alter the consequence of any defects that arise due to movement during cure. With the hourglass shape, defects located on the surface near the middle of the joint may have a more deleterious effect on performance than the defects would have on a square joint. However, to assess the impact of movement during cure, hourglass shape joints exposed to movement during cure should not be compared to square joints that are not subject to movement during cure.

## CONCLUSIONS

- For one tested sealant, the performance improvement created by switching from square to hourglass or shallow rectangular joints is not large enough to boost the ASTM C 719 rating from ± 25% to ± 50%. The computer analyses indicate some lesser degree of improvement is likely. Further testing is needed to pinpoint the exact degree of improvement.

- The actual degree of performance improvement is likely to depend heavily on the particular balance a sealant has between adhesive and cohesive strengths. Hourglass shapes tend to reduce the overall peak stresses and to shift them away from the bondline.

- Square test specimens are likely to minimize variability in test results and produce more conservative results than will shallower joints.

## ACKNOWLEDGEMENTS

Thanks to Simpson Gumpertz & Heger Inc. for their support of development of this paper.

## REFERENCES

[1] Tons, E., "A Theoretical Approach to Design of a Road Joint Seal," Highway Research Board Bulletin, No. 229, 1959, pp 20 - 53.

[2] Tons, E., "Geometry of Simple Joint Seals Under Strain," New Joint Sealants, Building Research Institute, Publication No. 1006, 1963, pp 40 - 61.

[3] Cook, J. P., "A Study of Polysulfide Sealants for Joints in Bridges," Highway Research Board Bulletin, No. 299, 1965.

[4] Cook, J. P., "Effects of Specimen Length on Laboratory Behavior of Sealants," Highway Research Record, No. 80, 1965, pp 74 - 78.

[5] Catsiff, E. H., Hoffman, R. F., Kowalewski, R. T., "Predicting Joint Sealant Performance of Elastomers by Computer Simulation - I. Justification of Method," Journal of Applied Polymer Science, Vol. 14, 1970, pp 1143 - 1158.

[6] Catsiff, E. H., Hoffman, R. F., Kowalewski, R. T., "Predicting Joint Sealant Performance of Elastomers by Computer Simulation - II. Results in Simple Extension and Compression," Journal of Applied Polymer Science, Vol. 14, 1970, pp 1159 - 1178.

[7] Catsiff, E. H., "Predicting Joint Sealant Performance of Elastomers by Computer Simulation - III. Simulation of Single- and Multi-Step Extension of a Stress-Relaxing Material," Journal of Applied Polymer Science, Vol. 15, 1971, pp 1021 - 1028.

[8] Myers, J. C., Behavior of Fillet Sealant Joints, Master's Thesis, Department of Materials Science and Engineering, Massachusetts Institute of Technology, Cambridge, MA, June 1989.

[9] Myers, J. C., "Behavior of Fillet Sealant Joints," Building Seals and Sealants: Material, Properties and Performance, ASTM STP 1069, Thomas F. O'Connor, Ed., American Society for Testing and Materials, Philadelphia, 1990.

[10] Private communication with James Lomax, Rohm and Haas Co., 1993.

Olga Yakub Berman[1]

# BUILDING SEALANTS IN RUSSIA

**REFERENCE:** Berman, O. Y., ''**Building Sealants in Russia,**'' Science and Technology of Building Seals, Sealants, Glazing, and Waterproofing: Fifth Volume, ASTM STP 1271, Michael A. Lacasse, Ed., American Society for Testing and Materials, 1996.

**ABSTRACT:** In this paper, the author presents information concerning building sealants in Russia.

The paper includes data of demand and provision of construction industry with sealants, the features of sealants application and comprehensive description of types and application areas of sealant materials.

Technical requirements and physical properties, current experimental data on performance capabilities of some types of sealants in Russia are presented and discussed.

Possibilities of organization of a joint sealants business in Russia are described in the conclusion.

**KEYWORDS:** building sealants, elastomeric sealants, non-hardening sealants, application features, artificial weathering, durability prediction

---

**INTRODUCTION:** New Russia is attracting much more attention and interest of engineering, scientific and business people and organizations. The unlimited industrial potential of the country, its rich natural resources and great demand for various products and services are worth serious consideration.

Russia has a highly developed industry, many branches of which are steady consumers of sealant materials. These main branches are:

1. Construction industry
   - a) Residential buildings
   - b) Commercial buildings
   - c) Industrial buildings
   - d) Highways
   - e) Bridges
   - f) Airports
   - g) Tunnels
   - h) Dams
   - i) Other construction

---

[1]Consultant, 2101 Winchester Ave., Suite B-12, Philadelphia, PA 19115

2. Automotive
3. Aircraft
4. Space industry

The major consumer of sealants in Russia is the construction industry. The demand for building sealants in Russia is great and it is less than half met. For instance, only in residential construction sub-branch the annual demand for sealants is over 45,000 metric tons and the production of elastomeric sealants never exceeded 7000 metric tons. So far as the country continues to build and repair houses (about 30 million sq. m. in new construction and up to 6 million sq. m. in reconstruction) sealant materials permanently remain at the top of the list of building materials Russia needs most.

Russia started the manufacturing of sealant materials much later than western countries and the production of these materials was constantly behind construction, the residential construction in particular. Residential construction had to make do with leftovers and off-grades from military and defense supply or use common caulks instead of elastomeric sealants. For special unique installations (such as high-rise buildings of Cosmos hotel and Presidium of the Academy of Science in Moscow) and repairs (such as Lenin Mausoleum in Moscow, Victory Monuments in Volgograd and Minsk) separate lots of sealants were purchased from foreign companies. During the last two decades several types of elastomeric building sealants (polysulfide, urethane, silicone and others) were developed and introduced in Russia. Some of them are comparable to the analogous sealants put on the International Market. But the annual production of these materials is far from enough and their quality is very unstable. The cause of the matter one can find in extreme deficiency and instability of raw materials and polymers as well as in underdeveloped technology and poor quality control. As a result, numerous failures in weather proofing of joints between structure elements in buildings occur. In climatic zones with severe weather conditions the percentage of leaking and frost penetrated joints between wall panels reaches 10% and 20% in windows. All this requires additional sealant materials and expenses for repairs. The demand for sealants used in repair works total up to 40% of the demand for sealants in new construction.

**The features of sealants application**

Along with the generally accepted requirements that contemporary sealants have to meet, there are also some very specific requirements because of the features of their application in construction.

The great majority of exterior joints in residential construction are joints with one-stage sealing. This puts sealants in the first and only line of defense against the weather. About 10% of all joints are drainable and less than 1% are joints with two-stage sealing. The main problem with this type of joints is frost penetration in the winter.

Construction works and joints seal installation in Russia goes on all year round, including winter. That is why sealants there have unusually wide service temperature range, - i.e. (-50) - 70°C or (-58) - 158°F and special application temperature limitations i.e. (-20) - 50°C or (-4) - 122°F. In winter time non-hardening sealants are applied preheated to the temperature 30 - 40°C (86 - 104°F). During application a portion of preheated sealant is kept in thermos-box on a traveling cradle.

After storage at room temperature, elastomeric sealants should be kept in thermos-boxes as well. The data on ultimate low application temperatures allowed for various sealants are given in column 6 of Table 1.

Russian elastomeric sealants for winter application don't cure at low (below 0°C) temperature. The sealants used in winter are formulated to be thixotropic and low modulus in the whole range of service temperature. That allows to install them in the joints at low temperature. The sealants stay in the joints and work as caulks until warm weather comes and then they begin to cure.

The great majority of exterior walls in building are precast concrete panels grey, white and brownish in color. That is why the color range of sealants is also limited. If a contractor needs special color, he specifies it in his order.

Russia is very large country with centralized supply system. Construction enterprises that carry out the new construction in big cities and regions order and receive sealants once or twice a year. The main source of transportation are rail ways. So the shelf life of elastomeric sealants is required to be at least 1 year and only for one component elastomeric sealants (which are not very popular there) it is allowed to be 6 months.

In Russian cities residential construction are mostly multi-story buildings with precast concrete wall panels from 3 to 8m long. The accuracy both in manufacturing and in erection of wall panels in usually not high. When plans call for 20±5mm wide joints, as constructed joints may vary from 10 to 40mm in width and even reach 50mm.

The description and application areas of sealants and some other sealing materials widely used for weatherproofing outer joints in buildings are given in Table 1.

Urethane sealants, which are prominent in western countries, are not being used in Russia yet, though R & D works are going on for a long time. Recently, the attention of sealants formulators there is also attracted to modified polyethers, such as silicone modified polyethers and thiokol modified polyethers. Several two and one-part sealants were developed and tested showing good results. The materials provide a winning combination of properties, good weathering characteristics and adhesion to a variety of substrates, good workability and curing characteristics even at the lowest temperatures. But these sealants are not into mass production.

## Requirements and Properties

There are several National Standards in Russia, specifying Classification, General Requirements, Test Methods and Application of Construction Sealants. These standards are common and compulsory for all manufacturers and users.

Russia participated in ISO Committee 59/ Subcommittee 8 and accepted general recommendations and principles of ISO standards in the national standards, though some test methods differ in test procedures, apparatus and in test conditions (in conformity with application features).

Table 2 shows Nomenclature of Characteristics and General Requirements that all elastomeric sealants have to meet according to National Standards in force, as well as technical data on two elastomeric sealants, produced in Russia.

Table 3 shows major application properties of the same two sealants in comparison with the similar commercially available elastomeric construction sealants, produced in USA, Finland and Germany. All sealants are multicomponent and recommended for application in joints between precast concrete wall panels and other exterior dynamically moving joints. Technical data taken from manufacturers data sheets.

Although test procedures used by manufacturers are somehow different, the comparison of the properties seems possible and significant for practical purposes.

Standard tests are used in quality assurance and product development as a reliable way to ensure customers satisfaction and to compare materials. But when it comes to durability prediction, there is no standard test method. In Russia the durability of sealants is charged in terms of deterioration of major properties during artificial weathering cycles or during service years. Service lives of

TABLE 1 -- Sealants and other materials for weatherproofing of joints between precast concrete wall panels in buildings.

| Product name (Trade Mark) National Standard or Specification | Material presentation (description) | Total Movement Capability | Uses, Application | Quantity estimation kg/lin.m (lb/lin ft.) | Notes |
|---|---|---|---|---|---|
| 1 | 2 | 3 | 4 | 5 | 6 |
| Buteprol<br><br>GOST 14791 | Non-hardening construction sealants on the base of butil, polyisobutylene, ethylene-popylene and isoprene rubbers.<br>One component, remains plastic, high viscous, ready to use caulk. One of the lowest priced caulks available on the market.<br>Colors: grey, brown | ± 10% | Sealants<br>Recommended for sealing exterior one stage and drainable joints between average sided precast concrete panels and for sealing joints between panel interfaces and window or door units. | 1, 05<br>(0,7) | Service temperature range:<br>(-50) – 70°C<br>(-58)÷ 158°F;<br>Application temperature limitations:<br>(-20) – 30°C<br>(-4) ÷ 85°F<br><br>Needs priming |

| | | | | | |
|---|---|---|---|---|---|
| Tegeron<br><br>TU 21-28-87 | Non-hardening construction sealant on the base of butil rubber and styrene.<br>One component, high viscous caulk with inclusions of styrene granules, remains plastic during application and service.<br>Colors ivory, grey | ± 10% | Recommended for sealing exterior one stage and drainable joints between average size precast concrete panels and for sealing joints between panel interfaces and window or door units in North regions of the country. | 0,7<br>(0,46) | Service temperature range:<br>(-60) ÷ 70°C<br>(-76) ÷ 158°F<br><br>Application temperature limitations:<br>(-25) ÷ 30°C<br>(-13) – 86°F<br><br>Needs priming |

| | | | | | |
|---|---|---|---|---|---|
| AM-0.5<br><br>TU 84-246 | Two-part nonsag, low-modulus chemically curing elastomeric polysulfide construction sealant.<br>Colors: grey, black | ± 25% | Recommended for sealing exterior one-stage and drainable moving joints between large side precast concrete panels as well as expansion joints between masonry metal and other units. | 0.35 (0.23) | Service temperature range:<br>(-50) – 70°C<br>(-58) – 158°F;<br><br>Application temperature limitations:<br>above 0 – 30°C<br>above 32 – 86°F<br><br>No priming is required |
| LT-1<br><br>TU 38.403484 | Two-part non-sag, low-modulus Laprotiol-polymer based elastomeric joint sealant LT-1 has excellent primerless adhesion to most common construction surfaces.<br>Colors: ivory, grey | ± 25% | Recommended for sealing exterior joints, subject to cyclic movement, between large side concrete panels, as well as masonry, metal curtain wall joints. Provides a durable, weathertight seal. | 0.35 (0.23) | Service temperature range:<br>(-50) – 70°C<br>(-58) – 158°F;<br><br>Application temperature limitations:<br>(-20) – 30°C<br>(-4) – 86°F<br><br>No priming is required. |

| | | | | | |
|---|---|---|---|---|---|
| Gerostom<br><br>TU 21-23 | One-part nonsag, medium modulus carboxyl terminal epoxy based elastomeric constructon sealant.<br>Colors: beige | ± 20% | Recommended for sealing exterior joints between moving precast concrete panels, perimeter caulking etc. The product is especially suited for application where rapid cure is required. | 0,4<br>(0.26) | Service temperature range:<br>(-40) ÷ 70°C<br>(-40) – 158°F;<br><br>Application temperature limitations:<br>(-20) – 40°C<br>(-4) ÷ 104°F<br><br>No priming is required |
| Germabutil-1<br>Germabutil-2<br><br>RST-5018 | Two-part butil rubber solvent based construction sealants.<br>Colors: beige, black | ± 15% | Recommended for sealing exterior joints between average size wall panels.<br>Germabutil-1 should be applied on dry surface,<br>Germabutil-2 both on dry and wet surfaces. | 0.3<br>(0.2) | Service temperature range:<br>(-50) – 80°C<br>(-58) ÷ 170°F;<br><br>Application temperature limitations:<br>(-20) – 30°C<br>(-4) – 86°F<br><br>Priming is required. |

| | | | | | |
|---|---|---|---|---|---|
| Elastoseal<br>11-06<br>TU 6-02-775 | One-component non-sag, low modulus elastomeric sealant on the base of silicone polymer.<br><br>Colors: white, transparent | ± 25% | Recommended for sealing interior and exterior joints between large size panels in severe climatic conditions. | 0,3 (0.2) | Service temperature range: (-55) - 90°C (-67) - 194°F;<br><br>Application temperature limitations: (-10) - 30°C (+14) - 86°F Priming recommended with concrete surfaces. |
| PRP<br>GOST 19177 | Porous closed cell neoprene rubber based vulcanized gaskets. Round diameter from 10 to 60mm. (from 0.4 to 2.4 in) and rectangular 30x40 and 40x60 mm (1.2x1.6 and 1.6x2.4in) cross sections. | - | Back-up-materials<br>Recommended for installation in interior and exterior joints to control the depth of sealant and to provide additional air tightness of joints. Round gaskets are used in vertical joints, rectangular - in horizontal joints. | 1.05 lin.m (3.46 lin. ft) | Service temperature range: (-30) - 70°C (-22) - 158°F; For frost resistant gaskets: (-60) - 50°C (-76) - 122°F Installation at temperature above (-20)°C (-4)°F Recommended compression in joints: 25-50% of initial dimention. |

| | | | | | |
|---|---|---|---|---|---|
| Vilaterm-SM<br><br>TU 6-05-221-872 | Closed-cell poly-ethylene foam gaskets.<br>Round diameter from 10 to 60mm (from 0.4 to 2.4 in), and rectangular 30x40 and 40x60 mm (1.2x1.6 and 1.6x2.4in) cross sections. | - | Recommended for installation in interior and exterior joints to control the depth of sealant and to provide additional air-tightness of joints. Round gaskets are used in vertical joints, rectangular - in horizontal joints. | 1.05 lin.m (3.46 lin. ft) | Service temperature range:<br>(-60) - 70°C<br>(-72) - 158°F.<br><br>Installation is recommended at temperatures above:(-10)°C (+14)°F.<br>Recommend compression in joints 25-50% of initial dimention. |
| Gervolent<br><br>TU 21-29-46 | Uncured rubber based tape 180mm (7.2 in) wide, 1.2 mm (0.05in) thick. | - | <u>Tapes and aprons</u><br>Recommended for lining air-tight insulation of vertical joints between wall panels. | to suit job require ments | Service temperature range:<br>(-40) - 80°C<br>(-40) - 176°F<br>Must be glued with adhesive.<br>Installation at temperature above (-10)°C<br>(-14)°F |

| | | | | | |
|---|---|---|---|---|---|
| Gerlen-D TU400-1-165 | Non-hardening self-sticked tape lined with synthetic fabric 100 (4in), 120 (4.8in) and 200 (8in) mm wide; 3mm (0.12in) thick | - | Recommended for lining air-tight insulation of vertical joints between wall panels. | to suit job require-ments | Service temperature range: (-50) - 60°C (-58) - 140°F. Sticks to the concrete surface after priming. |
| Rubber Sheet GOST 7338 | Sheet rubber, frost and weather resistant | - | Rubber sheets are used for water conducting aprons in drainable vertical joints. | 220 x 220mm (8.8x 8.8in) | Glue on adhesive to the upper edges of adjoining panels in the zone of intersection of vertical and horizontal joints. |
| Folgoiozol<br><br>GOST 20429 | Sheet of double layer material: thin chequered aluminum foil, covered from below with bituminous rubber binder. | - | Folgoizol is used for water conducting aprons in drainable vertical joints. | 220 x 220mm (8.8x 8.8in) | Glue on bituminous - rubber mastics |

TABLE 2 -- Requirements for Elastomeric Construction Sealants and their Properties

| **Nomenclature of Characteristics** [1] **GOST 4.224** | **General Requirements** [1] **GOST 25621** | **Technical Data** [2] **LT-1** [1] **TU 38.103650** | [2] **AM-0.5** [1] **TU 84-246** |
|---|---|---|---|
| 1 | 2 | 3 | 4 |
| General-duty characteristics | | | |
| •Application temperature limitations | -20–30°C (-4–86°F) | -25–30°C (-13–86°F) | 0÷30°C (32–86°F) |
| •Working life, hours | 2–24 | 12 | 2-4 |
| •Ultimate tensile strength | 0,1–0,7MPa (14,5–101,5 psi) | 0,25MPa (36,25psi) | 0,5MPa (72.5psi) |
| •Ultimate tensile elongation, % | min 200% | 400% | 250% |
| •Type of failure [3] (A, C or A-C) | only cohesion | cohesion | cohesion |
| •Ultimate tensile elongation at min. low temperature, % | min 0,5E at -40°C (-40°F) | 400% at -50°C (-58°F) | 170% at -50°C (-58°F) |
| •Movement capability | ±25% | ±25% | ±25% |
| •Resistance to flow at specified temperature | max 2mm at 70°C (0.08in at 158°F) | 2mm at 70°C (0.08 in at 158°F) | 2mm at 70°C (0.08 in at 158°F) |
| •Tensile adhesion at 100% extention | 0,03–0,06 MPa (5–10psi) | 0,05MPa(7.2psi) | 0,05MPa(7.2psi) |
| Shrinkage, % | max 5 | 4 | 5 |
| Hardness, Shore A | 20÷30 | 20 | 20 |

1) GOST & TU - National Standards and Specifications

2) Product name (Trade Mark)

3) A - adhesion; C - cohesion; A-C - adhesion & cohesion.

| | | | |
|---|---|---|---|
| •Cure time at 23°C (75°F) | 14 days | 14 days | 14 days |
| •Resistance to weathering | min 15 cycles | 20+ | 15+ |
| •Resistance to heat aging | should pass | passes | passes |
| •Resistance to water immersion | should pass | passes | passes |
| •Resistance to cyclic tension and compression, cycles | min 1000 | 2000 + | 1500 + |
| <u>Structural characteristics</u> | | | |
| •Density | 1000-1200kg/$m^3$ (62.45-74.94 lb/$ft^3$ | 1200kg/$m^3$ (74.94 lb/$ft^3$) | 1200kg/$m^3$ (74.94 lb/$ft^3$) |
| •Sealant quantity per lineal length of joint (at joint width 20mm (0.8in) | 0.3kg/lm (0.2lb/ft) | 0.35kg/lm (0.23lb/ft) | 0.35kg/lm (0.23lb/ft) |
| <u>Reliability characteristics</u> | | | |
| •Expecting service life (durability) | min 15 years | 20-25 | 15-20 |
| Shelf life (storage time) | min 12 months | 12 months | 12 months |
| •Storage conditions | Below 25°C(77°F) Unopened, dry area | Unopened, dry area, below 25°C(77°F) | Unopened, dry area, below 25°C(77°F) |

| | | | |
|---|---|---|---|
| <u>Transportability characteristics</u> | | | |
| •Gross mass of one package | Max 40kg (88.2lb) | 40kg (88.2lb); 5kg(11lb) | 40kg (88.2lb); 5kg(11lb) |
| •Packaging | | 40kg(88.2lb) pail 5kg(11lb) pail + 0.75kg(1.65lb)can | 40kg(88.2lb) pail 5kg(11lb) pail + 0.75kg(1.65lb)can |
| <u>Ergonomics characteristics</u> | | | |
| •Service temperature range | -40–70°C(-40–158°F) | -50÷70°C(-58–158°F) | -50–70°C(-58–158°F) |
| •Toxicity | Should Pass | Passes | Passes |
| <u>Aesthetic chracteristics</u> | | | |
| •Color(s) | White, grey, others upon request | ivory, grey, brown, others upon request | light grey, ohters upon request |
| •Staining | No stain or color change on the surface of white mortar plate | Passes | Passes |

**TABLE 3** -- Application Properties of Various Multicomponent Elastomeric Construction Sealants

| **Application Properties** | **LT-1 (Russia)** | **AM-0,5 (Russia)** | **Permapol ® RC-2 ("PRC," USA)** | **Rubber-seal-2K ("Sadolin," Finland)** | **Terostat-2K ("Teroson," Germany)** |
|---|---|---|---|---|---|
| Service Temperature Range °C (°F) | -50–70 (-58–158) | -50÷70 (-58÷158) | -40–82 (-40–180) | -40–90 (-40–194) | -40–80 (-40÷176) |
| Application Temperature Limits | -20–50 (-4)–122 | 5–30 (41–86) | no data | 5÷30 (41–86) | 5–30 (41–86) |
| Movement Capability, % | ±25 | ±25 | ±25 | ±25 | ±25 |
| Working Life | 12 h | 2-4 h | 8 h | 3-48 h | 2-4 h |
| Resistance to flow, vertical | max 2mm (0,08in) at 70°C (158°F) | max 2mm (0,08in) at 70°C (158°F) | max 2mm (0,12in) at 51°C (120°F) | max 2mm (0,08in) at 50°C (122°F) | max 2mm (0,08in) at 50°C (122°F) |
| Hardness, Shore A (Standard Conditions) | 20 | 30 | 30 | 30 | 30 |
| Tensile Strength MPa (psi) | 0.25 (36.25) | 0.5 (72.5) | 0.83 (120) | 0.7 (101.5) | 0.5 (72.5) |

| | | | | | |
|---|---|---|---|---|---|
| Ultimate Elongation, % | 400 | 250 | 600 | 300 | 300 |
| Shelf Life, mo | 12 | 12 | 12 | 12 | 12 |
| Colors | Ivory, Grey, Brown | Light Grey | White, Black, Bronze, Limestone, Pine, Others | White, Grey, Black, Brown | Black, Brown, Grey |
| Packaging | 40kg (88lb) metal pails; 5kg (11lb) plastic pails + 0.75kg (1,65lb) cans | 40kg (88lb) pails | 5.7l (1½gal) kits, 11.4l (3gal) pails | 2000ml (½gal) and 4000ml(1gal) cans | 2500ml (0.06gal) cans |

20 and even more years for some elastomeric sealants are predicted under a variety of environmental conditions and some correlation between degradation in artificial and in natural conditions was observed.

The results of artificial weathering of two mentioned above elastomeric sealants LT-1 and AM-0,5 are given in Table 4. For artificial weathering the test method and aparatus were similiar to those, described in ASTM C 793-91. The differences were in speciments (in this case - sealant between two standard mortar blocks) and in evaluation of the results (in this case - changes in tensile - strength properties, not only the description of tested speciments).

Table 5 shows the changes in tensile properties for the same two sealants after up to 5 years service life in exterior walls joints in continental climate.

**TABLE 4** -- Tensile - Strength Properties after Artificial Weathering

| | LT-1 | | | AM-0,5 | | |
|---|---|---|---|---|---|---|
| **Number of Cycles** | **Tensile Strength MPa x 10** | **Ultimate Elongation %** | **Type of failure (cohesion, adhesion)** | **Tensile strength MPa x 10** | **Ultimate Elongation %** | **Type of failure** |
| Initial Properties | 1.98 | 479 | cohesion | 5.99 | 827 | cohesion |
| 1 | 1.92 | 4.38 | cohesion | 5.81 | 643 | cohesion |
| 3 | 1.87 | 416 | cohesion | 5.67 | 506 | cohesion |
| 5 | 1.85 | 366 | cohesion | 5.11 | 500 | cohesion |
| 10 | 1.52 | 372 | cohesion | 5.12 | 405 | cohesion |
| 15 | 1.39 | 345 | cohesion | 5.76 | 364 | cohesion |
| 20 | 1.24 | 280 | cohesion | 6.53 | 240 | cohesion & adhesion cracks on surface |
| 25 | 1.05 | 231 | cohesion & adhesion, cracks on surface | --- | --- | --- |

**TABLE 5 --** Tensile - Strength Properties after up to 5 years Service Life.

| | LT-1 | | AM-0,5 | |
|---|---|---|---|---|
| Walls Orientation | Tensile Strength MPa x 10 | Ultimate Elongation % | Tensile Strength MPa x 10 | Ultimate Elongation % |
| Initial Properties | 1.98 | 479 | 5.99 | 827 |
| after 1 year | | | | |
| South | 1.96 | 485 | 6.50 | 644 |
| North | 1.83 | 518 | 5.98 | 713 |
| East | 2.07 | 492 | 6.68 | 686 |
| after 3 years | | | | |
| South | 1.94 | 416 | 7.13 | 508 |
| North | 1.67 | 489 | 5.11 | 516 |
| East | 2.21 | 457 | 6.97 | 499 |
| after 5 years | | | | |
| South | 1.02 | 334 | 6.98 | 450 |
| North | 1.62 | 396 | 5.76 | 498 |
| East | 2.19 | 443 | 6.54 | 434 |

**CONCLUSION:** Russia and the USA have very similar approach to R&D of construction sealants and in testing of these materials. At the same time there are essential differences in manufacturing and application of building sealants caused by different state of technology, construction technique, economical conditions and societal attitudes.

In light of this, the organization of a new sealant business in Russia seems to be timely and to have good prospects. There are two principal ways to carry out this business:

1. To negotiate direct contracts for selling building sealants to specific consumers.

2. To organize on the territory of Russia joint ventures (enterprises) to produce sealant materials and other products, and then to sell them in Russia and to other countries.

Both of these ways do not exclude but in fact supplement each other. However, the first way will achieve rather limited results and accordingly yield limited return. Because of the current situation in Russia (the extreme deficiency of converted currency), in this case only the purchase of separate lots of materials for special, unique installations can be expected.

The second way provides both interested sides (the USA and Russia) with ample opportunities because:

1. In Russia there are considerable free capacities suitable for the production of sealants.

2. Various raw materials and some polymers are also available.

3. There is an unlimited supply of labor especially in the regions where these free capacities are concentrated.

With rather limited investments it is possible to modernize manufacturing processes, develop formulations, train attending personnel, and organize delivery of inaccessible ingredients (if necessary).

In short, these joint ventures can start manufacturing and selling sealants to the countries of the former Soviet Union and to other countries (Eastern European countries, India, Syria, Egypt, for example). These countries possess converted currency and are interested in buying sealant materials of high quality.

# Joint Sealant Failures: Case Studies and Methods of Evaluation

Russell J. Kenney and Richard S. Piper[1]

# CASE STUDY OF SEALANT REVERSION

**REFERENCE:** Kenney, R. J., and Piper, R. S., **"Case Study of Sealant Reversion,"** Science and Technology of Building Seals, Sealants, Glazing, and Waterproofing: Fifth Volume, ASTM STP 1271, Michael A. Lacasse, Ed., American Society for Testing and Materials, 1996.

**ABSTRACT:** Numerous failures of building seals and sealants have logical and, in most cases, simple explanations as to why the seal or sealant failed to function as intended. This paper focuses on a failure mechanism that is not as apparent or simple to explain; failure caused by improper mixing of two part sealant or the reversion of cured sealant. On two separate buildings with the same sealant, the authors encountered a varied mode of failure. This paper documents the investigation of one of these projects and discuses the failure phenomena in relation to the orientation of the sealant joints and the analysis of physical properties of seals exposed to varied conditions of ultraviolet radiation and elevated temperatures. It is concluded that reversion in certain types of polyurethane sealants compounded with benzoate based plasticizer is more likely when exposed to conditions of high levels of ultraviolet radiation and elevated temperatures.

**KEYWORDS:** absorbency, EIFS, polyurethane sealant , sealants, sealant reversion, uncured sealant, urethane sealant

## INTRODUCTION

A high-rise building built between 1985 and 1987 and located in a major northeastern city utilized an exterior insulation and finish system (EIFS) exterior wall cladding with sealant joints between prefabricated EIFS panels and building seals at wall openings. Shortly after the building was occupied, water penetration was noted at window openings. Within a few years, a change in the properties of the sealant was noted and, by 1994, a substantial amount of the sealant had failed. Failure was considered a condition where the sealant was no longer capable of normal movement and had deteriorated to a point where the sealant was no longer an effective air and water seal.

[1]Director of Testing and Senior Architect, respectively, R. J. Kenney Associates, Inc., P. O. Box 1748, Plainville, MA 02762

This paper discusses the condition of the sealant, a laboratory test performed to determine the change in properties of the sealant, and the condition of the sealant relative to the ultraviolet exposure of the sealant joint or a combination of moisture and ultraviolet exposure.

## INVESTIGATION

The exterior wall investigation documented problems with the double-hung window system. It was found that prior to 1993, the window system caused water to be trapped behind the sill seal. This, in combination with sunlight and elevated temperatures, caused the sill seal to deteriorate (Fig. 8). It was also noted that no sill flashing was used, and consequently, no weeps were provided at the sill. The use of an open-cell joint backer material in some locations contributed to the problem.

The seventeen story structure was clad with a prefabricated EIFS panel system. The EIFS was recessed at the windows, and the window frame had a fin configuration (Fig. 1). The investigation found that the distance from the face of the recessed portion of the panel to the face of the window frame was not adequate for the application of an adequately sized seal.

The sealant was identified as a two part, low modulus, polyurethane sealant conforming to ASTM C 920 Standard Specification for Elastomeric Joint Sealants, Type M, Grade NS, Class 25.

### Window Openings

The EIFS panel design and the absence of a return at the edge of the window frame created a difficult joint to properly seal at the window perimeter. In most cases, the depth of the joint was such that no backer material other than bond breaker tape could be used. The joint depth was less than 3 mm in some cases. Manufacturers typically recommend a minimum depth of 6 mm for polyurethane sealants. No bond breaker tape was found during the investigation, and the sealant was typically applied directly against the back of the joint, creating three-sided bonding for the majority of the window perimeter. The sealant was often bonded only to the edge of the approximately 1.6 mm thick aluminum window frame, resulting in a very thin and inadequate bond line.

Proper joint design requires the sealant be applied onto the face of the window frame approximately 10 mm when there is inadequate depth of frame for a recessed joint. This provides adequate bond line width for good adhesion of the sealant to the window frame. It is also important that the sealant have adequate thickness (minimum 6 mm) over the window frame to prevent a cohesive failure of the sealant at the edge of

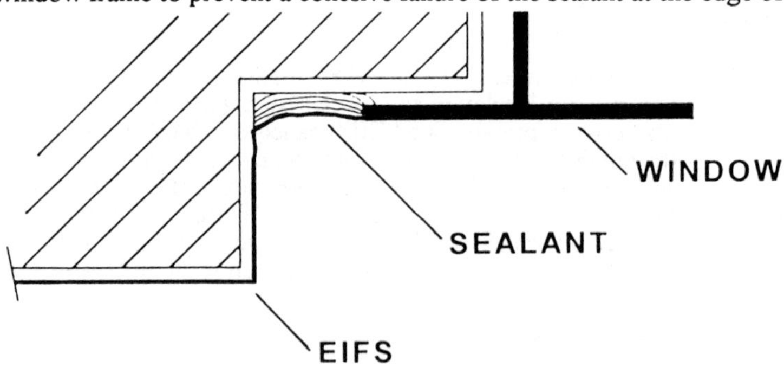


FIG. 1--Typical sealant joint at window jamb with 3-sided bonding and bond to narrow edge of window frame.

FIG. 2--Sealant is crazed and has failed in adhesion to the window frame.

the frame while allowing for weathering of the surface of the urethane. In Figure 2, the sealant is crazed and not consistent with the condition expected of a good quality two-part polyurethane sealant at the age and exposure of the sample.

Besides internal window leaks that allowed water to accumulate at the sill of the window behind the sealant, there were minor cracks in the EIFS at the corners of openings in a few locations. There were also a few cracks in the EIFS caused by EIFS base coat between pieces of insulation which allowed water penetration adjacent to the perimeter of the windows (Fig. 3).

FIG. 3--Gap between EPS boards caused crack in EIFS.

FIG. 4--Gummy, non-elastic sealant at EIFS panel joint.

Prefabricated Panel Joints

Sealant failures were found on both vertical and horizontal joints between the prefabricated EIFS panels. Most of these failures were due to the non-elastic properties of the sealant that had either been improperly mixed, improperly proportioned, or had reverted due to exposure to sunlight, moisture, and elevated temperatures.

Fig. 4 is a typical example of sealant that has been improperly mixed. The sealant was tacky, textured, and marbled. In some cases, the sealant had sagged out of the joint (Fig. 5). Although not all of the tacky, textured sealant was marbled, on the south

FIG. 5--Sealant sagging out of EIFS panel joint.

FIG. 6--Deteriorated sealant at vertical EIFS panel joint.

elevation in areas with the greatest exposure to the sunlight, both the vertical and horizontal joints at the perimeter of the panels were severely deteriorated.

Fig. 6 is a close-up view of severely deteriorated sealant in a vertical joint. Part of the EIFS was removed at this location to investigate the condition of the substrate and to determine if the gypsum sheathing had been damaged by exposure to moisture. No

FIG. 7--Reverted sealant at window jamb with high sun exposure.

damage to the gypsum sheathing was found, nor was there evidence of water infiltration. Hence, water infiltration was not considered, in this instance, to be a factor contributing to the premature deterioration of the sealant

Sealant Joint Exposure

Sealant samples were taken from all elevations and from areas on the same elevation with different exposures to sunlight. Samples were taken from the first three levels and at the roof level.

In general, it was observed that sealant in many locations appeared to be affected by exposure to sunlight. In the case of the sill joints, where elevated moisture conditions were known to have occurred, the sealant was severely deteriorated. At the window jambs, due to the recessed window design and building orientation, one jamb would often have much greater exposure to sunlight. It was noted that there was a substantial difference in the condition of the sealant, depending on the orientation of the joint, and that the joints with greater exposure to sunlight were consistently in poorer condition. The joints at the window heads were recessed and, therefor, not exposed to sunlight. These joints were not deteriorated and were consistent with the condition expected of six to seven year old urethane sealant joints.

Fig. 7 shows a typical surface and partial internal condition of sealant at a window jamb that had high sun exposure for more than 6 hours a day on average for at least six months of the year. The outer skin is crazed and slightly darker than the soft center of the joint. The interior color is uniform without a marbled effect, indicating that improper mixing was not a problem.

Sealant in joints under window that were exposed to both sunlight and high moisture levels were often severely deteriorated (Fig. 8). Note the deep crazing and deterioration of the joint. In addition to the reverted or uncured condition, the joint lacks a backer material and has inadequate lap (less than 3 mm) onto the window frame.

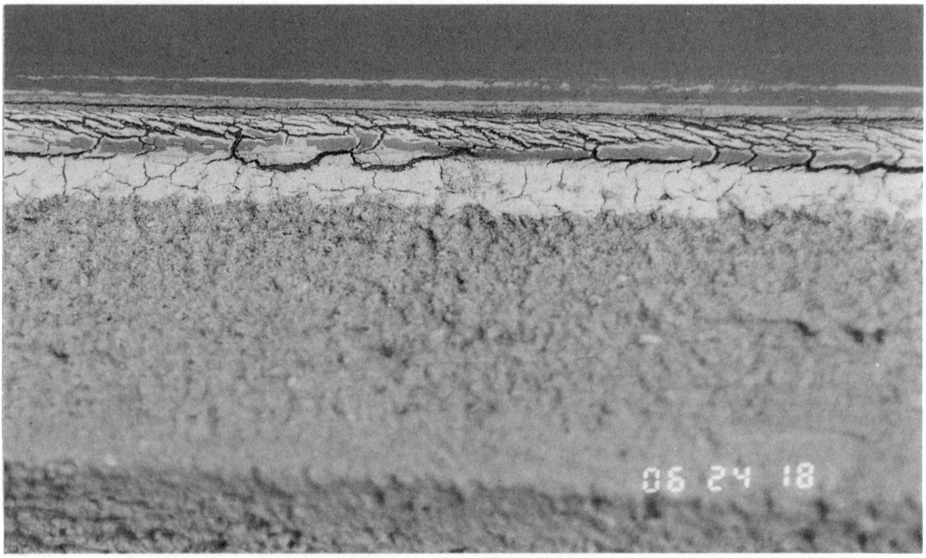


FIG. 8--Deteriorating sealant at window sill with deep crazing and poor bond to EIFS and Window frame.

TABLE 1 --Analysis of sealant joints by type and location

Vertical Panel to Panel Joints

| Elevation | Failed* | Non-reverted** |
|---|---|---|
| South (west) | 8 | 2 |
| South (east) | 5 | 6 |
| West (north) | 0 | 2 (shade) |
| West (south) | 3 | 2 |
| North | 5 | 6 |
| North (east) | 4 | 1 |
| East | 3 | 4 |

Horizontal Panel to Panel Joints

| Elevation | Failed* | Non-reverted** |
|---|---|---|
| South | 1 | 3 |
| South (east) | 4 | 0 |
| West (north) | 0 | 4 (shade) |
| West (south) | 5 | 4 |
| North | 1 | 9 (shade) |
| North (east) | 2 | 1 |
| East | 3 | 5 |

Window Jamb to EIFS Joints

| Elevation | Failed* | Non-reverted** |
|---|---|---|
| South (west) | 13 | 13 |
| South (east) | 10 | 0 |
| West (north) | 0 | 10 |
| West (south) | 2*** | 19 |
| North | 2 | 30 |
| North (east) | 6 | 6 |
| East | 4 | 16 |

* Failed was defined as uncured, non-elastic, or reverted sealant.

** Non-reverted indicates joints that were not uncured or were only partially reverted.

*** Improperly mixed sealant, either very hard or uncured with multiple colors.

## Analysis of Sealants by Type and Location

There are approximately 24,000 linear feet of EIFS panel joints and 22,000 linear feet of window perimeter joints. A field survey of the first three levels and the roof level was completed, accounting for approximately 23 percent of the total sealant used on this project.

In general, if the surface of the sealant had become tacky and non-elastic, the majority of the sealant was in that condition. Only the partially reverted joints required a decision on how to classify the failure mode. Those on the shade side of the window reveals were not affected by exposure and were easily classified. There was a definite division between the failed and acceptable sealant joints. The joints on the south elevation were without question the most severely deteriorated.

Approximately 28 percent of the window seals, not including the sill joints, had failed. Only seven percent had failed on the shade side of the building, while 63 percent had failed on the sun side of the building. Of the EIFS panel joints, 55 percent of the vertical joints and 38 percent of the horizontal joints had failed.

Table 1 summarizes the number of fully reverted, partially reverted, and non-reverted sealant joints observed in the field survey of the first three floors of the building.

## Laboratory Analysis

Sealant samples were taken and sent to a laboratory for testing to determine if there were any differences in properties between samples taken from areas of the building with different exposures to sunlight.

Test Method    Five grams of the sample were placed into a new, clean, glass centrifuge tube. Five ml of chloroform were added and mixed with the sample using a vortex mixer for one-half hour. After centrifuging, the chloroform extract was cast as a thin film onto a KRS-5 crystal. Analysis was then performed using Fourier transform infrared spectroscopy (FTIR). The absorbency of the peak at 455 $cm^{-1}$ was base line corrected from 600 to 412 $cm^{-1}$. The infrared spectra during the analysis was generated using a Bio-Rad FTS-7 FTIR spectrometer.

Test Results    Sample No. 4 was expected to have the lowest absorbency, because it was never exposed to direct sunlight on the building. Sample No. 12 had moderate exposure, and the test results were within the limits expected of a six to seven year old sealant subjected to some heat aging. Samples No. 10 and 16 are similar to the sealant shown in Figure 7 and were soft and tacky. The analysis of the sealant indicates the presence of absorbency peaks which are not consistent with reference spectra of normal, fully cured polyurethane sealants. The increasing intensity of these absorbencies correlates well with the increasing softness of the samples and the degree of sun exposure (Table 2).

TABLE 2 -- Absorbency and exposure.

| Sample No. | Sample Orientation | Absorbency at 455 $cm^{-1}$ | Ultraviolet Exposure |
|---|---|---|---|
| 16 | horz. | 1.172 | full exposure |
| 10 | vert. | 0.841 | full exposure |
| 11 | vert. | 0.170 | partial exposure |
| 12 | vert. | 0.091 | moderate exposure |
| 4 | vert. | 0.049 | shade |

The results are consistent with the field survey in that, on areas of the building where the sun was more directed at one side of the window than the other, the fully or partially exposed side of the window had a substantially higher percentage of soft and non-elastic sealant than did the less exposed side. This indicates the uncured or reverted condition was at least partially due to exposure.

Samples of six to eight year old sealant of the same type, grade, and class as the sealant used on this project had FTIR results using the same test method as employed on this project of under 0.250 absorbency at 455 $cm^{-1}$.

Samples were also tested for the presence of a benzoate compound (possibly a diproplyene glycol dibenzoate) that was once used as a plasticizer with these particular sealants. Part of the reason for requesting the FTIR test and of determining the possible presence of a particular plasticizer was the similarity between the deteriorated sealant on this project and on a project constructed at approximately the same time and known to have used the same sealant. Similar sealant reversion was found on both projects, and both projects used sealant with the same or similar plasticizer.

The relative concentration of urethane functionality in the suspect samples was compared to the control material and was found to be similar. This indicates that the cause of the gummy consistency of the sample was not improper mixing nor incomplete curing. The reaction products are present in the proper relative amounts, and it is not believed that the shelf life of this product was exceeded.

Additional Tests

Tests are proposed for the sealant on this building to determine if samples taken from the shaded or moderate sun areas are prone to revert to a gummy consistency similar to the joints on the sun exposed elevation of the building. Test would use the same FTIR techniques. There is a question whether the color of the sealant in combination with the plasticizer contributed to reversion of the sealant.

## SUMMARY

The project investigated had numerous areas where the sealant was improperly applied, having joint configurations not consistent with industry standards, including inadequate lap onto abutting surfaces and three-sided bonding These conditions are common problems, and failures due to these defects are easily explained.

Window leaks at the corner joints allowed water to accumulate behind the sill seals. This, combined with exposure to sunlight, caused or contributed to the sealant reversion at the window sills.

On joints other than under the window sills, bulk property changes due to thermal aging and the degree and depth of the deterioration encountered could not be explained solely by thermal aging and the availability of moisture. Some other effect was a factor in the premature failure of the sealant.

It was evident form the survey and test samples of the vertical window joints and EIFS panel joints that some failures in which moisture was not evident were due to the orientation of the building and hence exposure to higher levels of sunlight and consequent elevated temperatures. For example, it was observed that at windows with one jamb exposed to significantly more sunlight than the opposite jamb, the sealant at the more exposed jamb generally had a greater degree of softening. This occurred in areas where the sealant had been applied at the same time by the same mechanic using the same batch of sealant for both jambs.

The window head joints are recessed and consequently not exposed to sunlight, and the sealant did not soften.

The condition of the gypsum wall sheathing was investigated in a number of locations for evidence of water damage which would indicate possible exposure to elevated moisture levels which, in combination with sunlight, could have contributed to the softening or reversion of the sealant. No evidence of elevated moisture levels was found at the vertical or horizontal panel joints nor at the window head and jamb joints.

## CONCLUSION

- The building is located in a major northeastern city with numerous buildings with building seals that have had similar exposure and , in many cases, contain the same type, grade, and class of sealant. While these sealants show the effect of aging due to exposure, they have not deteriorated in the same manner or degree as the sealant joints on the building in question.
- The failure of the window sill sealant joints subjected to sunlight and elevated moisture conditions, although severe, is understandably a combination of elevated temperatures and wet conditions and is commonly accepted as a cause of deterioration, including reversion, of polyurethane sealant.
- The difference in the condition and performance of the sealant when comparing similar joints where water accumulation was not evident, with the sun exposure being the only variable, clearly indicates that exposure to sunlight alone influenced the sealant performance.
- The results of the FTIR analysis and physical condition of the sealant exposed to the most sunlight was substantially different than similar joints not exposed to the same degree of sunlight.
- The results of FTIR analysis of similar type, grade, and class of sealant of comparative age and exposure were generally less than 0.250 absorbency at 455 $cm^{-1}$. The soft gummy non-elastic sealant had substantially different values.
- The reversion due to exposure to sunlight of a two-part polyurethane sealant the full depth of the joint is not common nor a typical mode of failure and is substantially different than the 1 to 2 mm reversion known to occur with some polyurethane sealants.
- One other project built during the same time frame, using the same sealant, exhibited the same types of failures. In both cases, presence of a benzoate compound was detected, and the sealant was also the same color.
- Based on this investigation, it appears that the sealant on this project, compounded with a benzoate plasticizer, is sensitive to sunlight (elevated temperatures), and this resulted in reversion the full depth of the sealant thickness in less than seven years on elevations subjected to the most sunlight and elevated temperatures.
- Moisture and vapor conditions and the use of open cell backer material were considered as possible factors in the premature deterioration of the sealant. While these may be contributing factors, it is the authors' conclusion that the sealant experienced substantial change in properties due primarily to the effects of sunlight.

Michael J. Scheffler[1], and James D. Connolly[2]

# HISTORY OF BUILDING JOINT SEALANTS

---

**REFERENCE:** Scheffler, M. J., and Connolly, J. D., **"History of Building Joint Sealants,"** Science and Technology of Building Seals, Sealants, Glazing and Waterproofing: Fifth Volume, ASTM STP 1271, Michael A. Lacasse, Ed., American Society for Testing and Materials, 1996.

**ABSTRACT:** The use of sealants in facades became more widespread in the 1950s and 1960s with the increasing use of curtain wall construction. While most sealants specified for new construction today are silicone or urethane-based, older buildings may contain polysulfide and acrylic-based sealants.

For investigators trying to identify sealants used on older buildings, it is often useful to know which sealants were in common use when the building was constructed, or when the building was repaired. This paper briefly discusses and reviews the history of the use of commonly used building sealants over the last fifty years, detailing when specific sealant types were first developed, when they came into widespread use, or fell out of favor.

Historically, sealant selection has depended upon many factors such as the type of substrate, movement requirements, prior sealants used, cost and aesthetics. Improvements in performance capabilities of sealants and innovations such as structural glazing are discussed.

Some sealants have been found to contain hazardous substances. For example, some polysulfide sealants applied during the 1960s contain polychlorinated biphenyls (PCBs). Issues related to hazardous materials in sealants will be discussed as well as the performance consequences of the current change to sealant products to make them less harmful to the environment.

**KEYWORDS:** sealants, history, silicone, urethane, polysulfide, acrylic, latex

---

[1]Consultant, Wiss, Janney, Elstner Associates, Inc. (WJE), Chicago, Illinois, USA.

[2]Senior Consultant and Manager of the Erlin, Hime Associates Division of Wiss, Janney, Elstner Associates, Inc. (WJE), Northbrook, Illinois, USA.

Modern building joints are typically sealed with elastomeric sealant materials that adhere to joints in walls to form a barrier against air and water, and remain flexible when exposed to the weather. Along with principal cladding components, elastomeric sealants in building facades are a key element in defense against water penetration and air infiltration into the building. This paper briefly discusses and reviews the history of the use of commonly used building sealants over the last fifty years.

Often during the investigation of existing sealants in buildings, investigators are asked to determine the type of sealant present in order to properly recommend repairs. It can be helpful to the investigator to know when certain types of sealant were available to the building construction industry in order to make a preliminary determination. In order to accurately determine the type of sealant present, laboratory testing is often required.

The four principal characteristics of any of these sealants are adhesion, cohesion, elasticity of the sealant, and weathering ability. It is important in the identification and of sealant to understand these characteristics.

- Adhesion is the property that describes the sealant's ability to bond to the adjacent substrate.
- Cohesion is a measure of the integrity of the body of the sealant; it describes the sealant's ability to resist cracking and tearing within itself.
- Elasticity is the property by which the sealant accommodates joint movement without adhesive or cohesive failure or permanent deformation.
- Weathering ability is the property of the sealant to retain its physical characteristics when exposed to the environment.

Elastomeric joint sealants when tested for adhesion and cohesion under cyclic movement are rated to withstand an increase and decrease of the original joint width. The joint movement capabilities of high performance elastomeric sealants are generally greater than +/-25 percent of joint width.

Sealants are usually placed in joints as a viscous liquid that is either pourable for horizontal applications or non-sagging for vertical surface joints. These materials cure after placement in the joint, taking on the configuration of the joint. Recently, pre-cured pre-formed sealants have become available for use in building construction.

Sealants can be either single-component, not requiring mixing, or multi-component, requiring mixing 2 or 3 components just prior to application.

Elastomeric joint sealants are intended to be installed on clean, dry surfaces and may require the application of a primer to the substrate surface in order to promote adhesion. Joint preparation often includes the installation of sealant back-up materials to control the shape of the sealant installed.

A variety of elastomeric sealants are used today in building construction to seal joints between various building elements. They include the more commonly used and well known sealant types such as silicone, urethane, and polysulfide sealants typically used on commercial and high-rise buildings. Other types more typically used on small-scale and residential buildings include acrylic, latex, and butyl-based products. The following is a brief description and history of each of these sealant types:

## Oil and Resin-Based Caulks

Oil-based caulks, which were being used as early as the turn of the century, were often termed "putty." Oil-based caulks typically are made primarily with calcium carbonate and a drying oil such as linseed, fish, soybean, tung, or castor oil [1].

Oil-based caulks, including those manufactured prior to the 1950s have relatively limited movement capability of +/- 2 percent to +/- 5 percent; hence they cannot be considered elastomeric sealants. Prior to the 1950s, sealed joints typically were incorporated into load-bearing masonry construction that had joints that experienced little movement. Although the movement capability of these sealants was limited, they usually proved to be adequate for sealing joints in construction of the first half of this century, particularly when painted regularly. Essentially, these products were the only ones available for use in building construction in the sealing of building joints up to about 1950.

Oil-based caulks have better weathering characteristics when painted. Typically, the joint material and adjacent substrates were painted with an oil-based coating. Since the life expectancy of oil-based caulks is about 3 to 5 years, many older buildings that originally had joints sealed with these kinds of caulk have been resealed with newer, better performing sealants.

## Polysulfide Sealant

With the increased popularity of modern curtain wall construction in the 1950s, the demand increased for a sealant joint material that could seal joints experiencing greater movement. The first of the elastomeric joint sealants that became widely used was polysulfide sealant which became available to the construction market in the early 1950s.

Polysulfide sealant is a high performance sealant that is based on a synthetic polysulfide polymer or rubber. Polysulfide polymer is a mercaptan terminated, long-chain aliphatic containing disulfide linkages. The first polysulfide polymer was developed by the Thiokol Chemical Corporation in 1929. The polysulfide sealants available in the 1950s were based on the Thiokol polysulfide polymer, which dominated the market. These polysulfide sealants consisted of two-component products contained in separate containers. The components were mixed together just prior to application. One of the components consisted of the Thiokol rubber polymer, filler, and plasticizer, while the other consisted of lead dioxide paste and plasticizer. In the 1950s the cost of polysulfide sealant was approximately 10 times that of the more traditionally used oil-based caulks, preventing to a large extent the increase in use of this sealant. Increased use of polysulfide sealants increased as performance became better known and as labor costs steadily increased relative to material costs over the next 20 years.

In 1961, one-component polysulfide sealants that did not require mixing became available for use. Among the early manufacturers of a one-part polysulfide sealant were Dicks-Armstrong-Pontis, Inc. (DAP) and A. C. Horn Companies (Horn).

Shortly after their introduction it was noticed that some formulations of polysulfide sealant caused a pink colored stain on adjacent concrete surfaces. This staining was found to be related to the adhesion additive, an extractable water-soluble phenolic resin that turned pink in an alkaline environment. This problem was overcome by manufacturers in the early 1960s through reformulation of their products [2].

The widespread use of polysulfide-based sealants grew rapidly to the point where during the 1970s approximately 35 different formulations of polysulfide sealant based upon the Thiokol polymer were available [3]. These formulations, made by various manufacturers, were performance tested and approved for use by Thiokol.

Thiokol eventually ceased providing control and approval for various proprietary polysulfide sealants. The use of polysulfide sealants diminished significantly after the 1970s, as polysulfide sealant performance became more erratic. Competition from excellent urethane and silicone sealants became strong. However, many buildings still have joints containing polysulfide sealant since the service life of some polysulfide sealant under certain conditions can exceed 20 years. Polysulfide sealant also remains in some joints that are covered by newer applications of other types of sealant.

In 1985 polysulfide sealants accounted for approximately 5 percent of the sealant used in construction [4]. Today polysulfide sealants are still used for submerged joint applications because of superior resistance to water and chemical degradation and to a limited extent they are used for cladding joints. Polysulfide sealants typically have a movement capability of +/- 25 percent. Two-part polysulfide sealants do not exhibit physical properties that are as good as some urethane or silicone sealants, which have movement capabilities of +/- 50 percent, or greater.

Insulating glass windows manufactured for commercial and home use require high performance sealants to accommodate the movement of glass and aluminum and provide the low water vapor transmission rates required to prevent the windows from fogging. Elastomeric sealants that are stable and flexible over a wide temperature range are used for this application. Polysulfide sealants meet these requirements and are used in more than half of the insulating glass units manufactured [5].

## Silicone Sealant

The development of the silicone polymer used in silicone sealants dates from the early 1800s. In the United States, since the 1930s, silicone polymer has been developed for use as a sealant by the Dow Corning Corporation and by the General Electric Corporation. The first silicone sealants were two-component products developed in the early 1950s [6].

The silicone polymer is based on alternating silicon-oxygen atoms. Silicone sealants typically consist of a high percentage of the silicone polymer and may contain fumed silica, calcium carbonate, ground quartz, carbon black, talcs, plasticizers and other miscellaneous compounds [7]. Silicone sealants are high performance sealants that became available to the building construction industry around 1960 [8]; the first was manufactured by Dow Corning [9]. In 1963 another silicone sealant for construction use, manufactured by General Electric, became available. Both of these sealants were one-component products that did not require mixing. Silicone sealants applied at building sites have been primarily one-component products that are cured by moisture. Acetic acid is a curing by-product of certain silicone sealants. Other silicone sealants produce neutral by-products upon cure and are preferred for some building substrates. For example, acetoxy cured sealants are not recommended for building substrates that are acid sensitive such as concrete, mortar, limestone, and marble.

Silicone-based sealants are most frequently used for non-porous surfaces in high-movement joints. They are commonly used for metal and glass cladding substrates. Staining of porous substrates and dirt accumulation have been an ongoing problem for some silicone sealants.

Silicone sealants as originally produced had movement capability of +/- 50 percent. Today the movement capability of silicone sealants varies depending on type but can be as high as +100/-50 percent [10]. Silicone sealants typically considered to have better resistance to ultraviolet radiation UV, and heat aging, and better moisture stability than other types of sealants.

Silicone sealant was used for two-sided structural sealant glazing and to eliminate vertical mullions at store fronts during the 1960s [11]. In a two-sided structural sealant glazing joint, sealant is typically placed in vertical joints between glazing units in place of exterior structural mullion support components. The glass units are held in place along the top and bottom edges with mechanical supports. Four-sided structural sealant glazing, where sealant is the sole structural support of the glass and no exterior mechanical components are used, was first used in 1970 [12]. Silicone sealants are currently the only sealant used in structural sealant glazing, due to their ability to provide strong, durable bonds to glass.

Silicones used only in conjunction with polyisobutylene, which forms the primary moisture barrier, account for about one-fifth of sealants used in insulating glass unit manufacture.

Today silicone sealants are the most used sealant in the construction industry. At the time they were introduced, silicone sealants were more expensive than other sealants. The cost of silicone sealants from the 1960s through the 1980s was about twice that of other sealants. Silicone sealant today is still slightly more expensive than other types of sealant, but with material cost a small percentage of the total cost of installing sealant, sealant material cost is becoming less significant an issue.

Silicone sealants are among the best performing sealants available today.

## Urethane Sealant

Urethane sealants are based on polyurethane synthetic rubber which was first developed in Germany in 1937 and later in England in the 1940s [13]. In the 1960s the first sealants became available that were based on the urethane rubber. Besides the urethane polymer, urethane sealants also usually include fillers, colorants, plasticizers, thixotropic agents, adhesion-promoting additives, and solvents. Urethane sealant is based on a polymeric material prepared by the isocyanurate addition polymerization reaction between di- or polyisocyanurates and di- or polyfunctional hydroxyl compounds. These sealants were first typically used in joints exposed to abrasive action such as traffic areas. By the early 1970s urethane sealants were available from various manufacturers such as Pecora Chemical Corporation, Gibson-Homans Company, Tremco Manufacturing Company and others. The first urethane sealants available were two- or three-component products that required mixing just prior to application.

Single-component moisture-curing urethanes became readily available in the mid to late 1970s. One-component urethanes, in contrast to multi-component urethanes, cure by reacting with moisture in the air. Some are made with "moisture scavengers" such as calcium oxide, which serve to inhibit set until placement occurs.

Urethane-based sealants are high-performance sealants most frequently used for cladding joints and in particular, for porous surfaces in high-movement joints. Urethane sealants as originally produced typically had a movement capability of +/- 25 percent. Beginning in the late 1980s the movement capability of some urethane sealants was as high as +/- 50 percent.

## Acrylic Sealant

Acrylic sealants are based on acrylic polymer technology that dates to 1843. The first production of the polymer took place in the 1920s [14]. Acrylic polymers are based on thermoplastic resins formed by polymerization of the esters of acrylic acid. Typically acrylic sealants are used in small-scale construction because they have limited movement capability of not more than +/- 12.5 percent, and are therefore only appropriate for low-movement joints.

Acrylic sealants first became readily available to the building construction industry in the early 1960s. The first acrylic sealant was developed by the Tremco Manufacturing Company in 1958. Although their use peaked in the mid-1970s, acrylic sealants are still used today primarily for small scale low-rise construction.

Acrylic polymer based sealants typically are one-component solvent-curing products that include fillers, such as calcium carbonate and silica, and solvents, catalysts, thixotropic agents and plasticizers.

## Latex Sealant

Latex sealants were developed in the 1950s and typically are based on acrylic, styrene-acrylic, vinyl acrylic, polyvinyl acetate, and styrene butalene. These sealants are based on a colloidal suspension of rubber resin (synthetic or natural) in water. Latex sealants became available to the construction industry in the mid-1960s. Latex sealants are one-component and have a limited movement capacity of up to +/- 7.5 percent. Typical components of a latex sealant include water, ground calcium carbonate, plasticizers, small amounts of mineral spirits, ethylene glycol, surfactants, and pigments.

Latex sealants are typically used in light building construction and residential construction because of their limited movement capability. Latex sealants can be painted with latex paint, and are generally less expensive and easier to clean up than the high performance sealants. For the most part latex sealants have replaced oil and resin-based sealants in small building and residential construction.

Plasticized acrylic latex sealants became available in the mid-1980s [15]. Plasticized acrylic latex sealants have improved movement capability over the older original latex sealants and have movement capability of +/- 25 percent. These newer latex sealants have increased levels of plasticizer, but still have essentially have the same weathering and adhesion characteristics as the older original latex sealants.

Butyl Sealant

Sealants that are butyl rubber-based became available to the building construction industry in the mid-1950s, even though butyl rubber was being manufactured earlier in the century for other uses [16]. Butyl rubber is based on the polyisobutylene polymer, which is manufactured from gaseous hydrocarbons, and also contains a small percent of isoprene. Butyl rubber was developed to replace the natural rubber which was in great demand just prior to and during the Second World War. Isobutylene based sealants are typically used for glazing joints and splice seals in window units.

Butyl rubber based sealants are one-component and have a movement capacity of up to +/- 7.5 percent. Butyl rubber based sealants may include in their composition talc powder, calcium carbonate filler, polybutylene, mineral spirits, and adhesion promoters, as well as other compounds. Butyl-based sealants in general do not have the high performance properties of silicone, urethane, and polysulfide based sealants but do have special characteristics that make them useful for certain glazing applications. Butyl hot melts are used in insulating glass manufacture as a primary seal, since they have a much lower vapor transmission rate than silicone.

Joint Back-up Materials

To help performance, some oil-based sealant manufacturers recommended that oakum or equivalent filler be installed in joints to control the depth and amount of material in the joint. However, the history of joint fillers for sealant prior to the 1950s is not well documented.

Joint back-up materials have typically been recommended by manufacturers of sealants since the late 1950s. Since then, it has been recommended that the depth of the sealant in joints be controlled with foam backer rod fillers made of polyethylene or polyurethane, or polyethylene bond breaker tape.

Closed-cell backer rod has the advantage of not absorbing and transmitting water that leaks into the joint, thereby exposing the sealant to continuous periods of moisture contact. Open-cell backer rod has the advantage of being more easily installed because of its compressibility.

Some sealants formed bubbles while curing because of air or gases that were released from certain closed cell back-up materials cut during backer installation. To help eliminate this problem, in the 1980s, a new type of backer rod was developed. This rod is open cell foam with a continuous skin on its surface.

Staining

An often-observed problem, particularly with silicone sealants, is staining of adjacent substrate materials. Some silicone and polyurethane polymers contain certain types and amounts of plasticizers or oily polymers that increase the probability of staining. Staining can be evidenced by dirt accumulation or discoloration on the adjacent substrate due to migration of the plasticers or incompletely polymerized polymers into the adjacent substrate. Light-colored stone and masonry substrates are particularly vulnerable to migration and staining from certain silicone sealants. Laboratory studies can be performed to determine the fluid contents of silicones. The amount and viscosity of the silicone fluid in a silicone sealant plays on important role in the staining characteristics of particular proprietary products. Silicones that contain more than about 13 percent fluid stain vulnerable stones. Rain water running over incompletely cured silicone sealant and onto windows can leave tenacious spots on the windows.

Staining of paints or elastomeric coatings can occur when they are applied over a crack sealed with sealant, due to outward migration of the plasticizer from the sealant bead into the coating. Where light-colored sealants are applied over neoprene gaskets, the oil in the gasket may also stain the sealants.

Other Sealant Problems

Historically, all kinds of sealants have failed when the substrate hasn't been properly primed or cleaned. They have also failed progressively because water got into the joint and behind certain sealants, in absorbent back-up material such as open-cell backer rod. Many sealants still do not have good long-term adhesion if continuously wet, or in standing water for extended periods. Adhesive testing should be done under both dry and wet conditions when selecting a sealant.

When silicone sealants have been found to be tacky and fluid after curing, it was typically due to application of a batch of sealant beyond its recommended shelf life.

Problems with multi-component sealants, particularly urethanes, have often been related to incomplete or improper mixing. If mixing is incomplete, portions of the sealant will be fluid or uncured. Incomplete curing can also be the result of using materials beyond their shelf life, or from using components from inadequately sealed containers, or under certain circumstances, from improper formulation.

Curing at temperatures above 120 degrees Fahrenheit has made certain urethanes, the authors have observed vulnerable to "reversion" or premature degradation from ultraviolet light and moisture. Under ambient air temperatures of 80 to 90 degrees Fahrenheit, dark-colored substrates exposed to direct sunlight can reach temperatures exceeding 160 degrees Fahrenheit. Both cure and durability of sealants are affected by high temperatures. Some urethane sealants cured at high temperatures may later become fluid and bubbled to a depth of 1/8 inch, and gummy at lower levels. Several studies have been performed on this issue in the last few years as this problem has been recently recognized.

Under high temperature and high humidity curing conditions some urethanes, particularly one component urethanes, have developed excessive bubbling and blistering. One-component urethanes made alkaline with moisture scavenger particles that are too large and are exposed on the surface will react with rainwater and can cause stains on windows.

A rapid cure urethane sealant was developed in the late 1980s for prefabricated wall panel construction. This sealant did not cure properly and was withdrawn shortly thereafter from the building construction market.

Hazardous Sealant Components

A wide variety of materials have been used in sealant formulations. Sometimes these may include unusual, but potentially hazardous components. Some sealants including oil-based sealants and glazing compounds were formulated with lead [17]. Asbestos fiber was used as a component of at least one oil-based sealant that was available in 1961 [18]. Reviewer questions whether sealants manufactured today contain lead?

A variety of companies manufactured polysulfide sealants, and because of this formulations varied. For example some polysulfide sealants installed during the 1960s contained polychlorinated biphenyls (PCBs) as a plasticizer. Today, PCB's are a suspected health hazard. Therefore, special handling and disposal procedures are required when handling these materials. Since it is not possible to tell from visual inspection whether these chemicals are present in the existing sealants, basic hygiene and protection precautions should be taken. For example, gloves should be worn when handling sealant, and workers should wash their hands immediately after contact. Care should also be taken to prevent sealant containing hazardous chemicals from becoming airborne during removal operations. The presence of hazardous compounds may also limit available preparation techniques. For example, where PCBs are present, it may not be possible to grind out existing joints.

Some sealants today are being reformulated to comply with new regulations on materials in order to make them less harmful to workers and the environment. For example some urethane sealants have been recently reformulated to remove mercury as a curing catalyst. Although reformulated products may be sold under the same trade name as previous products, their performance may be substantially different. The performance consequences of these changes are as yet unknown and need to be further evaluated and monitored.

Sealant performance characteristics such as joint movement capability and durability have increased dramatically since the 1950s and 1960s. But we expect and demand more from sealants today. We are sometimes also interfacing some high performance products with residue of older generations of sealant. During the investigation of existing sealants in buildings it is beneficial to know about the histories of the various sealants typically used in order to properly and safely inspect, specify and install new sealant. A proper investigation of existing building conditions and sealants is important for successfully repairing or replacing high performance building joint sealants.

## REFERENCES

[1] Panek, Julian R. and John Philip Cook. Construction Sealants and Adhesives, third edition. New York: John Wiley and Sons, Inc., 1991, pages 181-184.

[2] Panek and Cook 1991, page 76.

[3] Panek and Cook 1991, pages 107-117.

[4] Skeist, Irving. Handbook of Adhesives, third edition. New York, New York: Van Nostrand Reinhold, 1990.

[5] Skeist 1990.

[6] Panek and Cook, Construction Sealants and Adhesives, third edition, pages 118-128.

[7] Panek and Cook 1991.

[8] Panek and Cook 1991.

[9] A. Jazujen, in Plastic Mortars, Sealants and Caulking Compounds, edited by R. B. Seymour, ACS Symposium Series 113. Washington, D.C.: American Chemical Society, 1979, pages 113-127.

[10] J.M. Klosowski and G.A.L. Gant. The Chemistry of Silicone Room Temperature Vulcanizing Sealants, in Plastic Mortars, Sealants and Caulking Compounds. Washington, D.C.: ACS Symposium Series 113, American Chemical Society, 1979, pages 82-83.

[11] Panek and Cook, Construction Sealants and Adhesives, third edition, pages 118-128.

[12] Panek and Cook 1991, pages 146-157.

[13] Panek and Cook 1991, pages 129-137.

[14] Panek and Cook 1991, pages 159-163.

[15] Panek and Cook 1991, pages 170-173.

[16] Panek and Cook 1991, pages 175-180.

[17] "Elastic" glazing compound manufactured by H.B. Fred Kuhls, from Sweets Catalog, 1944. D-P white lead putty manufactured by Dicks-Point Company, Dayton, Ohio, from Sweets Catalog, 1950.

[18] Asbestos fiber is described as a component of Vulcatex elastic caulking compound, manufactured by the A. C Horn Companies as described in their product literature, copyrighted in 1960. Product literature obtained from Sweet's Catalog, 1961.

# Structural Silicone Glazing: Design Evaluations and Applications

Voytek (W.) S. Gutowski,[1] Peter Lalas,[2] and Anthony P. Cerra[1]

# STRUCTURAL SILICONE IN CURTAIN WALLS

**REFERENCE:** Gutowski, W.S., Lalas, P., and Cerra, A.P., ''**Structural Silicones in Curtain Walls,**'' Science and Technology of Building Seals, Sealants, Glazing, and Waterproofing: Fifth Volume, ASTM STP 1271. Michael A. Lacasse, Ed., American Society for Testing and Materials, 1996.

**ABSTRACT:** Modern building facades are frequently constructed using Structural Silicone Glazing (SSG) technology. In the USA and South-East Asia, 45 and 65% of facades, respectively, are erected with the use of this technique, in which external cladding panels are attached to the facade framing by an elastomeric silicone adhesive (sealant). The chemical structure of silicone sealants assures chemical bonding across the sealant/substrate interface. Therefore, the bond integrity should be assured for at least the life of the building. The reliance of the facade integrity on adhesive bonding, however, and the relatively young age of the technology require caution and a thorough understanding of material performance and the influence of the service environment on the material properties. Often, hydro-thermal and loading factors are inadequately considered in the facade design, construction and maintenance. Also, the influence of the sealant joint geometry on the material performance is neglected. This paper addresses major issues concerning the above factors, as related to the structural design and maintenance of structurally glazed buildings.

**KEYWORDS:** silicone sealants, loads, thermal movements, bead dimensions, performance criteria, environmental stability, performance monitoring

[1]Senior Principal Research Scientist and Experimental Scientist respectively, CSIRO Div. Bldg, Constr. & Engineering, Melbourne, Australia. Postal address: P.O. Box 56, Highett, Victoria 3190, Australia.

[2]Director, Arup Facade Engineering, 477 Kent Street, Sydney, N.S.W. 2001, Australia.

## INTRODUCTION

The main function of structural silicone sealants is to permanently bond external cladding panels to the facade frame, and to ensure the structural integrity of the building envelope throughout its entire service life. The secondary purpose is to provide water tightness for the building.

A building envelope undergoes a range of static and dynamic movements as its components respond to changing temperature and wind conditions, seismic movements and various permanent deformations of the building frame. In response to all these movements, sealants must exhibit satisfactory adhesion and acceptable mechanical and rheological properties. From a sealant performance viewpoint the most important properties are the strength and elastic elongation. Both of these must satisfy the expected confidence level at the required safety factor defined as follows:

(a) strength safety factor

$$\text{safety factor} = \frac{\text{strength of the system}}{\text{all stresses from all sources}}, \qquad (1)$$

(b) safety factor with respect to the sealant elongation

$$\text{safety factor} = \frac{\text{total joint movement}}{\text{allowable sealant elongation}}. \qquad (2)$$

The strength of the system must be significantly greater than the design stress which is commonly accepted at the level of 0.138 MPa (or 20 p.s.i.). The sealant elongation capability typically is within the range of ±40 to ±100% and has to be greater than the design strain which is commonly accepted at the level of ±15%.

Silicone sealants are polymeric materials. As such they exhibit the well known dependence of mechanical and rheological properties with temperature and deformation (strain) rate. All material properties are also expected by the authors to be significantly dependent on the geometry of the adhesive bead.

To date, very little and frequently no data, are available for structural design purposes which consider the combined influence of temperature, deformation (strain) rate and geometry on the structural properties of elastomeric silicone sealants and self-adhesive tapes. None of these factors should noticeably influence the strength of adhesion, because this is usually greater than the cohesive strength of the sealant.

Therefore, the purpose of this work is to fill the existing knowledge gap by achieving the following goals:

(i) to determine the cohesive and rheological properties of silicone adhesives in order to provide meaningful and reliable data for structural engineers and sealant manufacturers,

(ii) to provide information on the physical properties of sealants for most typical sealant bead geometries,

(iii) to provide foundations for the subsequent development of codes of practice and standards in the use of structural silicone sealants and adhesive tapes in structural glazing.

## BASIC TERMS REGARDING SEALANT BEAD DIMENSIONS

Figure 1 provides an explanation of all the terms which relate to the geometry of sealant bead in structural curtain wall applications.

Another term used in the investigations and further analysis is the sealant bead aspect ratio which is defined as follows:

$$\text{Aspect ratio} = \frac{\text{bite}}{\text{glueline}} . \qquad (3)$$

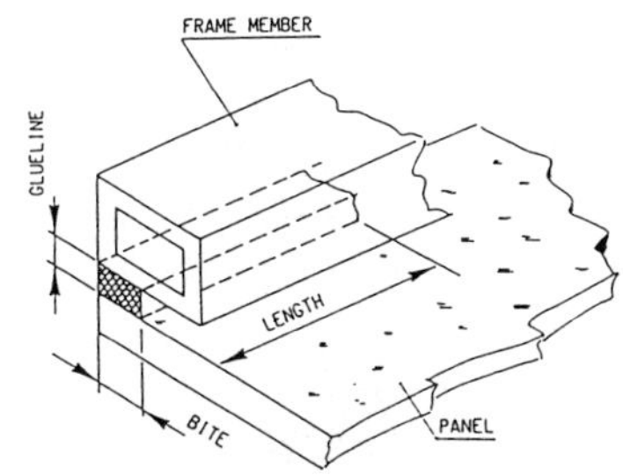


FIG. 1--Explanation of basic terms regarding dimensions of a sealant bead.

## TYPICAL LOADS AND JOINT MOVEMENTS

Building facades undergo a range of static and dynamic movements and loads which are typically of the following origin:

(a) Thermal movement
(b) Wind load
(c) Rotational deflection of panels due to wind pressure
(d) Dead load movement
(e) Creep, shrinkage, expansion and elastic deformation of the structural components.

### Wind Load

Wind load on the lee side of a building creates a zone of negative pressure which provides a negative force resulting in a sealant or adhesive tape being under a tensile force. The magnitude of this force depends on the wind speed, building location and orientation, building height and external shape, surrounding topology including adjacent building, and other factors. Figure 2 illustrates an example of negative pressure distribution on the lee side of a high rise building, as determined by model testing in a wind tunnel [1].

Wind load is the only direct loading (apart from dead load) which is always considered when a structural silicone joint is designed. Even so, wind load is considered to be a simple suction on the lee side only, evenly distributed over the facade panel it retains. Positive pressure from the wind is ignored.

Sealant manufacturers still only advise design of structural silicone joints using the trapezoid rule [1], ie.

$$b = \frac{\frac{B}{2} \times P}{F} + T, \qquad (4)$$

where: b - bite, B - short side of rectangular panel, P - design wind pressure (suction), F - design stress of sealant = 139 kPa (20 psi), T - tolerance

This rule, embodied by Eqn.(4) ignores the reverse loading effect (ie. compressive stress) that occurs at the corners, which is known about from plate theory [2], and the work done by Joe Minor [3]. A pressure in a thin plate will cause an upward force at the corners. This force coincides with the position of the weakest silicone joint. The corners are the positions where the installer stops and starts during sealant gunning, which results in the most poorly tooled joint details and the sealant bead is more than likely to have voids.

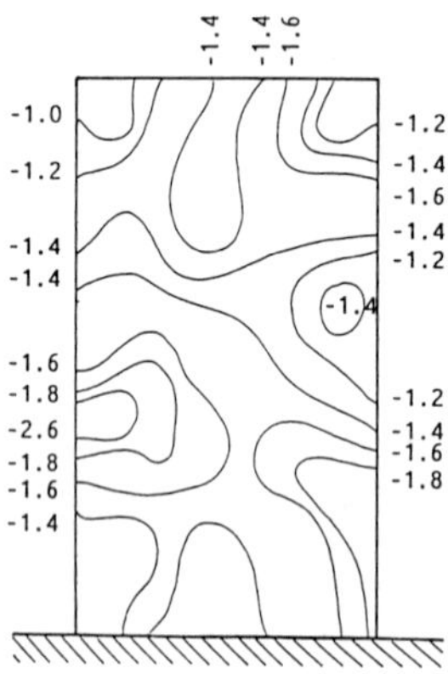


FIG. 2--Example of peak pressure isobars on the lee side of a building, as determined by model testing in a wind tunnel.

Sealant manufacturers continue to advise that the design stress for structural silicone to be used in Eqn (4), is 139 kPa (20 psi). There continues to be little justification for this nominal value. In accepting this nominal value, other factors which affect sealant joint strength, such as geometry are ignored.

Dead Load

It is recommended in structural glazing that the only structural loads imposed on a sealant bead are those affected by negative wind pressure. Accordingly, panels which are structural silicone glazed are usually supported on dead load (setting) blocks. However, sealant manufacturers advise that some dead load may be acceptable for sealant beads attaching single panels, assuming that the design stress value of this does not exceed 6.9 kPa (1 psi).

Double glazed units, however, must always be supported on rubber setting blocks. This is because the silicone bite would have to be unpractically large to keep the dead load design stress below 6.9 kPa. Also, the asymmetric load on the unit joint would pre-clude a warranty from a double glazed unit manufacturer.

Therefore typically only 6mm glass panels sometimes have their dead load supported by structural silicone alone without setting blocks. However, structural glazing without mechanical dead load supports is not recommended by these authors.

The effect of creep of structural silicone is not understood. Nor is the effect of joint geometry on joint long-term shear capacity.

Combination Loading

The simplest combination loading occurs often in shopfronts and sometimes in curtain walls. Figure 3 shows an example of a simple 90° joint between two pieces of glass which subsequently creates conditions for combination loading.

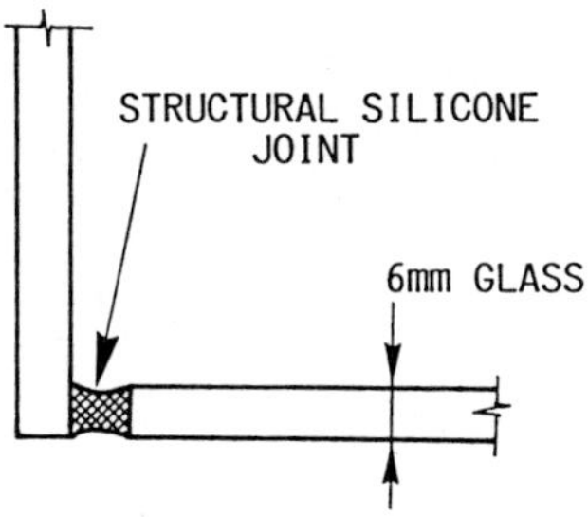


FIG. 3--Simple glass corner joint.

Negative wind load on each of the glass panels together with thermal movement (expansion and/or contraction) load the silicone joint with a combination of shear and tensile or compressive stress.

The sealant manufacturers give no guidance on how to design such a joint. Such a joint was found in a curtain wall in Sydney [4]. The panel width was 1200mm wide and the wind load 2.0 kPa. The sealant manufacturer's formula (trapezoidal rule, see Eqn (4)) requires a silicone bite of 8.7mm, in each direction. The joint size, however, was 6mm x 6mm, as governed by the 6mm glass thickness. The joint was undersized by a factor of 3, if the sealant manufacturer's advice, as above, was to be adhered to.

Figure 4 shows a corner joint in a double glazed unit. No written guidance can be found, again, on how to design either the double-glazed unit joints or the silicone joints holding the units together.

Another example of combination loading is dead load plus wind load. The sealant manufacturers are not clear on whether the allowable design stress is 20 psi for wind plus 1 psi for dead load, making a total of 21 or is the total load allowed to be 20 ie. 19 + 1?.

The problem of combination loading has been addressed by Sandberg and Ahlborn [5]. However, the applicability of their solution seems to be limited in case of building sealants because the equations proposed are satisfied for Hookean materials only. Building sealants satisfy this requirement at small strains only due to the known non-linearity of stress-strain characteristics of the majority of elastomeric materials.

It is clear to us that all direct loadings and secondary effects should be added, if occurring in combination, in any structural design. That is, the wind, dead load, temperature and moisture effects, ageing and secondary structural effects should be accounted for in joint design. However, a significant amount of research is required in order to fully understand and account for all these factors. Currently, their effect is compensated by an excessively high value of the safety factor which in structural glazing in Australia is taken as 5, while in the USA the value of 2.5 or 3 is deemed acceptable. By using a design stress of 139 kPa an ultimate safety factor of 4.5 to 5.5 is achieved in laboratory tests (carried out at the test rate of 50 mm/min).

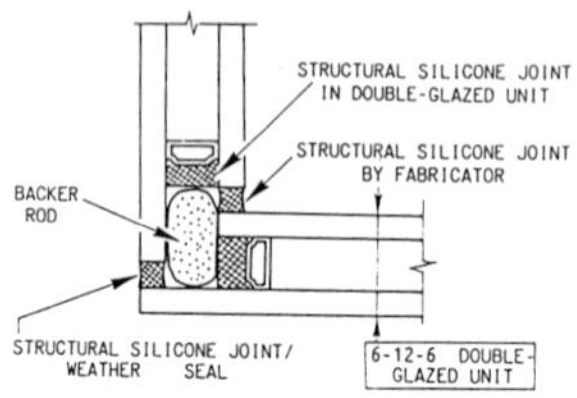


FIG. 4--Double-glazed unit corner joint.

Thermal Movements

All facade elements undergo significant thermal movements as a result of exposure to ambient temperature fluctuations and solar radiation. There are two, superimposed, components for the air temperature changes, as illustrated in Figure 5 and 6, eg.

(a) daily fluctuations, and
(b) annual fluctuations

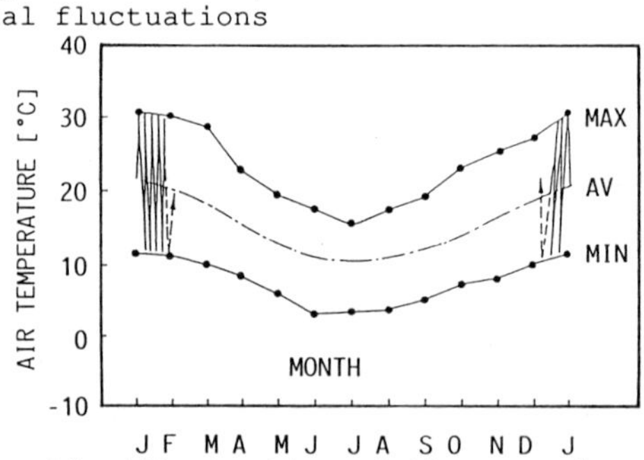


FIG. 5--A schematic illustration of the superimposed daily and annual changes in air temperature (typical for southern hemisphere) [6].

The presence of these two thermally induced movements has significant engineering ramifications due to the fact that all polymeric materials, including silicone sealants exhibit noticeable dependence of strength and strain at failure on the actual deformation rate.

For the sake of argument we assume that a silicone sealant, as prepared at a given constant geometry has been tested at 5 different rates, viz 0.01; 0.1; 1.0; 10 and 100 mm/min. The stress-strain curves obtained during the sealant testing are as illustrated in Figure 7.

In order to fully assess the significance of the example shown in Figure 7, one has to realise that as for any other materials, the mechanical properties of silicone sealants are determined in reference to standardised procedures. Accordingly, the cohesive strength and extension at failure of silicones, in accordance with the ASTM C1135 [7] are determined at the cross-head speed (test rate) of 10 mm/min. The question therefore arises, how this arbitrary test rate relates to the real rates of sealant deformation, occurring in real joints in the building facade. These real movement rates can be

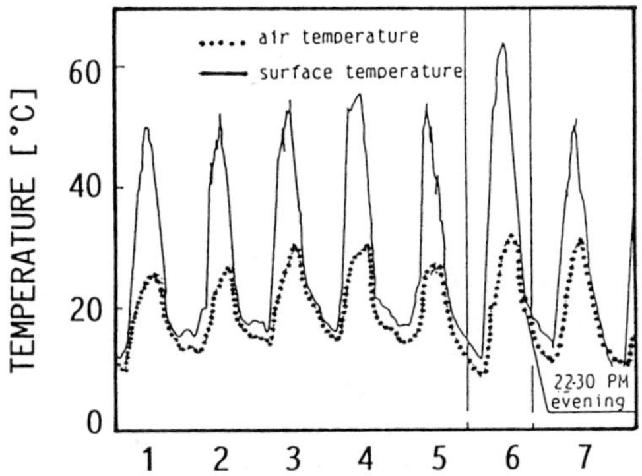


FIG. 6--Surface turf temperatures (assumed to be similar to a green reflective glass panel) recorded on the Parliament House in Canberra in the period 1-7 January 1986 [6].

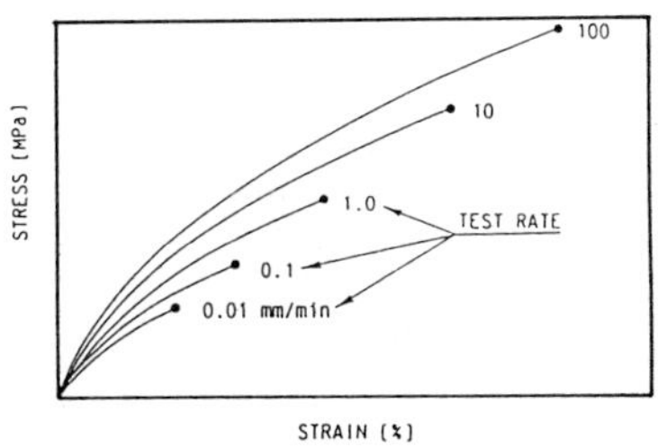


FIG. 7--An example of a typical pattern of stress-strain curves for a silicone sealant tested at 5 different rates (see also Ref 6)

easily estimated by relating the actual daily and annual thermal movements of facade components to the duration of these movements ie. thermal extensions and shrinkages, in accordance to the following formula:

$$V = \frac{\Delta L}{\Delta t} \qquad (5)$$

where V is the deformation rate; ΔL is the thermal expansion/shrinkage and Δt is the duration of the thermal expansion/shrinkage cycle.

In accordance with Figures 5 and 6, the duration of daily and annual thermal cycles is

(i) for daily temperature fluctuations: 8 to 10 hours (480 to 600 minutes)

(ii) for annual temperature fluctuations: about 6 months ($2.628 \times 10^5$ minutes).

The above temperature changes result in the cyclic shrinkage and extension of wall panels causing significant variations in the width of weather seals, which undergo cyclic tensile-compression movement, as well as cyclic shear movement of the structural sealant bead.

The overall thermal movement ($\Delta L_{th}$) of the facade components is calculated as follows:

$$\Delta L_{th} = L \alpha \Delta T_m \tag{6}$$

where L is the length (or height) of the facade element, $\alpha$ is the coefficient of thermal expansion of the facade material (see Table 1 for examples) and $\Delta T_m$ is the maximum temperature fluctuation. Although the values of thermal expansion coefficients are available in the tables it must be remembered that if composite elements are

TABLE 1--Coefficients of thermal expansion of materials $\alpha$ [$\times 10^{-6}$ mm/mm panel length $\times$ °C].

| Material | $\alpha$ |
|---|---|
| Glass plate | 9.3 |
| Aluminium panel | 23.3 |
| Cement board | 5.0 |

used for facade construction, the relevant value of $\alpha$ must be determined experimentally in order to obtain a realistic figure which often cannot be easily estimated. The value of $\Delta T_m$ in Equation (6) is calculated as a difference between the maximum ($T_{max}$) and minimum ($T_{min}$) wall surface temperature ie.

$$\Delta T_m = T_{max} - T_{min} \, . \tag{7}$$

Table 2 below provides data on typical temperature variations in major Australian cities, which is used to calculate the annual thermal movements in building facades. Table 3 provides examples of average daily and annual movement rates calculated for 2500mm long panels made of three types of possible cladding materials ie glass, aluminium and compressed cement boards, calculated for Canberra and Ottawa (Canada) [6].

An analysis of the discussion as above leads to the following conclusions:

(i) the maximum expected daily and annual thermal movements must be estimated and the greater value should be used for determining the optimum joint size.

(ii) the sealant strength and elongation (for a given joint geometry) used for structural design must be relevant to the actual values of these properties as exhibited by the sealant undergoing deformations at the rates (speeds) relevant to the daily and annual deformation rates resulting from daily and annual temperature fluctuations. The actual panel temperatures and not ambient temperatures should be used.

Rotational Deflection of the Facade Panel

This phenomenon is created by two combined factors whose influence is illustrated in Figure 8.

(i) thermal movement of the facade panel, and
(ii) negative wind pressure

The movement of the sealant is calculated (see Figure 8) as follows:

$$y = \sqrt{z^2 + x^2} \ . \tag{8}$$

TABLE 2--The highest and lowest monthly mean temperatures for major Australian cities.

| City | Temperature (°C) | |
|---|---|---|
| | $T_{max}$ | $T_{min}$ |
| Canberra | 31.7 (Jan) | -3.2 (July) |
| Melbourne | 31.6 (Jan) | 3.3 (June) |
| Sydney | 29.1 (Feb) | 4.3 (July) |
| Alice Springs | 39.1 (Jan) | -0.8 (July) |
| Perth | 36.3 (Jan) | 4.9 (Aug) |
| Rockhampton | 34.7 (Jan) | 5.1 (July) |
| Adelaide | 31.6 (Jan) | 4.1 (June) |
| City | Max. daily temp. changes (°C) (monthly average) | |
| Canberra | 24.3 (Nov) | |
| Melbourne | 19.3 (Jan) | |
| Sydney | 15.9 (Sept) | |
| Alice Springs | 25.1 (May) | |
| Perth | 22.4 (Feb) | |
| Rockhampton | 20.1 (Aug) | |
| Adelaide | 12.4 (Jan) | |

TABLE 3--Estimated average daily and annual thermal movement parameters for Australia and Canada (2500 mm long panels)

| | Daily movements | |
|---|---|---|
| Material | Extension (mm) | Speed (mm/min) |
| glass | 1.3 | $2.3 - 3.7 \times 10^{-3}$ |
| aluminimum | 3.3 | $6.0 - 9 \times 10^{-3}$ |
| cement board | 0.7 | $1.3 - 2 \times 10^{-3}$ |
| | Annual movements | |
| Material | Extension (mm) | Speed (mm/min) |
| glass | 0.8 - 1.6 | $3 - 6 \times 10^{-6}$ |
| aluminimum | 2.0 - 3.9 | $8 - 15 \times 10^{-6}$ |
| cement board | 0.45 - 0.83 | $2 - 3 \times 10^{-6}$ |

With regard to rotation due to wind effects, it is usually required that the panel (eg. glass) deflection in the centre is less than L/180 where L is the length of a panel. With regard to typical panel sizes this translates into about 0.5 to 1° rotation. However, this small rotation angle may add up to 3-5% of an extra movement to a typical sealant joint.

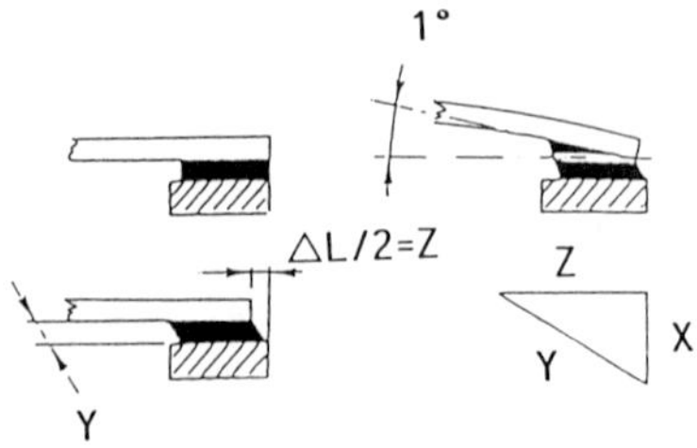


FIG. 8--Influence of wind pressure and thermal movement of the facade element on the rotational deformation in the sealant bead.

**DETERMINATION OF SEALANT PERFORMANCE CRITERIA**

The simplest way of determining sealant performance in terms of strength or extension at failure becomes available by presenting the experimental data "stress at failure v test rate" or "extension (strain) at failure v test rate", as illustrated in Figure 9. The test rate [mm/min] in Figure 9 is equivalent to the rate (speed) at which a sealant bead is deformed and stressed (in tensile or compression for weather seals and in shear for structural seals) upon exposure to the daily and annual fluctuations of the external panel temperature.

An example regarding the procedure for determining sealant failure criteria, as given below, has been presented for a 2500 mm long wall panel (Note: All data on strength and test rates are taken from a publication of K. Karpati [8]).

(a) type of sealant: silicone (undisclosed brand)
(b) sealant bead dimensions: 12.5 x 12.5 mm
(c) test rates [mm/min]: 0.0005; 0.002; 0.005; 0.01; 0.1, 0.5; 1.0; 5; 10; 50
(d) test temperature: 23°C

Figure 10 below illustrates the experimental relationship between the tensile strength and test rate (the latter presented on a logarithmic scale), prepared using a set of the re-interpreted data from Ref [4]. The test rate regions relevant to both, daily and annual deformation rates (see Table 3) are incorporated into the graph in order to provide a meaningful interpretation of the sealant properties (strength) for the joint design.

It can be seen from the graph that the experimental data correlating sealant bead strength with the test rate can be approximated by a straight line across the whole range of test deformation rates from 0.0005 to 50 mm/min. The abovementioned linear relationship includes the rates relevant to the daily temperature fluctuations (0.0023 mm/min for glass- and 0.0065 mm/min for aluminium panel - both 2500 mm long). The proposed pattern of the relationship conforms to the known studies on the time-temperature dependence of strength of rubbery materials (see Ref 9,10).

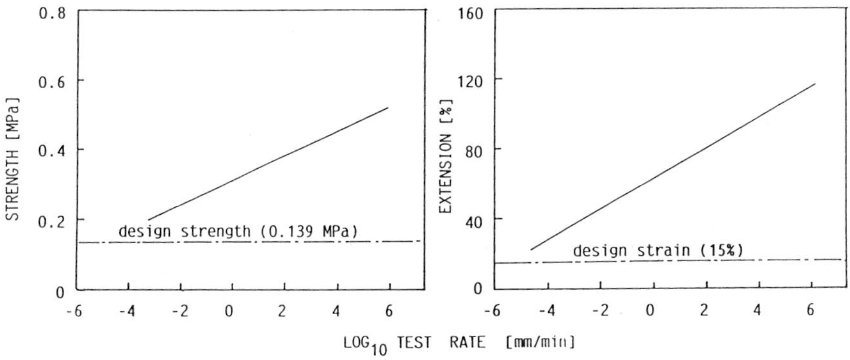


FIG. 9--Typical relationship between stress at failure or strain and test rate for determining sealant failure criteria in terms of stress at failure (strength), or strain (extension) at failure[6].

It appears from Figure 10 that the strength of the sealant at the rates relevant to the daily wall panel movements is about 4 to 5 times greater than the 0.138 MPa (20 psi) design limit for structural sealants. Also, the strength of this material at the deformation rates related to the annual temperature fluctuations seems to be above the required minimum.

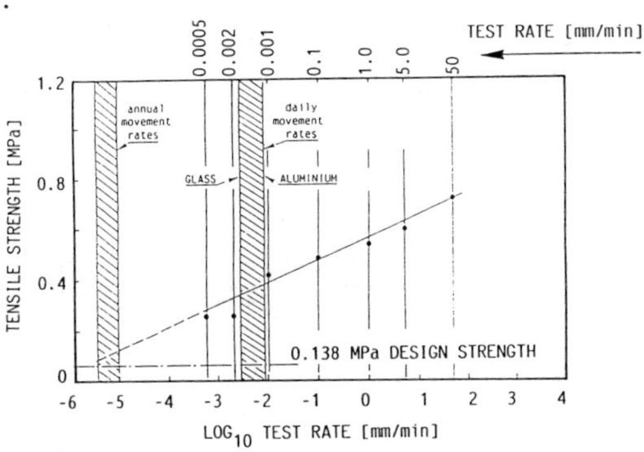


FIG 10--A plot for determining the sealant performance from the strength viewpoint.

## INFLUENCE OF SEALANT BEAD GEOMETRY ON STRENGTH AND EXTENSION CAPABILITY

The two main geometrical factors regarding sealants are the bite and glueline (the latter frequently denoted as joint width - typically in weather seals where the width is the gap between two adjacent wall panels), as defined in Figure 1.

As explained earlier, the sealant bite is designed to provide sufficient load-bearing capacity under the negative wind pressure, whilst the glueline (or joint width) is determined to accommodate all movements of external wall panels in relation to the supporting facade frame.

Until now, it has been assumed by designers that, like in all other engineering materials, the strength and strain at failure are intrinsic materials properties and as such are independent on the joint geometry. It is shown by the authors, however, that this assumption has to be revised. Figure 11 below illustrates the influence of the bead geometry on sealant strength and extension.

It can be observed from Figure 11 that both the sealant strength (ie. load-bearing capacity) and extension capability are significantly reduced by an increase of both glueline thickness and bite. The decrease of sealant extension capability appears to be more noticeable than the strength reduction. If, for instance, it is assumed that the strength of a 6 x 6 mm bead is 100%, the following can be seen for an example concerning a typical structural silicone sealant.

- doubling the bite from 6 to 12 mm reduced strength by 10%, and tripling the bite from 6 to 18 mm reduces strength by 20%,

- for the 1:1 aspect ratio, a glueline increase from 6 to 13 mm (sealant beads 6 x 6 and 12 x 13 respectively) results in a significant strength reduction as the bead size increases

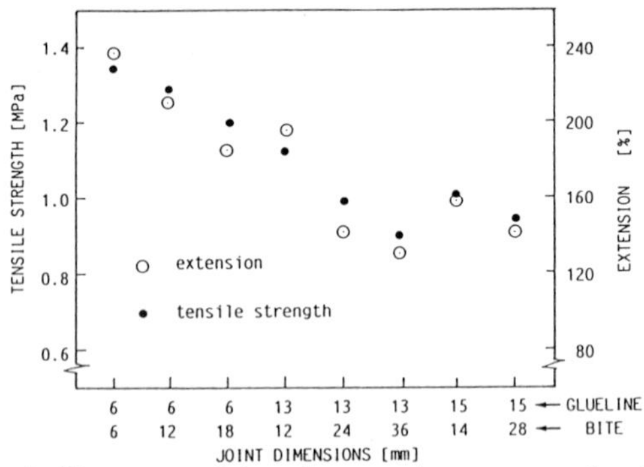


FIG. 11--The influence of bead geometry on sealant strength and extension [6]. Bead geometries, as indicated; temperature: ±20°C; test rate: 50 mm/min

A detailed study is underway to translate these observations into useful engineering terms.

## ENVIRONMENTAL STABILITY OF SILICONE SEALANTS

### Environmental Degrading Factors

Generally silicone sealants have good long-term weatherability [11]. Their resistance to oxidation, UV radiation, and ozone attack is superior to other sealants mainly due to the nature of their basic chemical structure.

The resistance of silicone sealants to water damage is a much more complicated issue. In this respect silicones behave rather paradoxically in that they are water repellents, but have high water vapour permeability which is important for the cure of these materials. The nature and behaviour of the bonding of silicone to substrates is still the subject of much research. It is thought that while the individual siloxane bonds are disrupted by the ingress of water at the interface, they are able to reform when the bondline dries out. For continuous underwater applications, silicones have to be specially formulated and used in conjunction with primers. Nevertheless, it is generally accepted that moisture is the main degrading factor.

Durability Testing

The numerous US and European standards were originally developed to assess the behaviour of elastomeric sealants when applied as sealants only and not as load-bearing adhesives. However, in the absence of more pertinent tests, these techniques have been accepted for the evaluation of SSG systems [12]. The strategy has been to precondition the specimens in a relatively severe environment before conducting a sensitive property measurement. For example, the major SSG sealant suppliers in the USA have specified the ASTM C794 Peel Test [13] on specimens that have been conditioned for one week in water at room temperature. Alternatively the Europeans have preferred a tensile test after a preconditioning regime involving both hot air exposure at 100°C and water at room temperature.

It is generally accepted that the peel test is only suitable for the qualitative assessment of failure mode and that a tensile test is necessary to determine the actual bond strength quantitatively. It is significant that ASTM has very recently published specification C1135-Test Method for Determining Tensile-Adhesion Properties of Structural Silicone Sealants [5]. Another important difference is that the French allow for two days drying time after water immersion, whereas ASTM specifies a maximum of only 10 minutes delay between the removal of specimen from water and testing. Considering the dynamic nature of the silicone/substrate bond in the presence of water, this discrepancy could prove to be very important.

Some of the authors have developed and investigated a novel test protocol which involves the application of a constant load or strain to a standard tensile specimen during immersion in water at elevated temperature [14]. It is perceived that its advantages to combine the critical parameters of load, temperature, moisture and exposure time within the one test. By stressing the bondline, failure is often induced during the exposure, otherwise the standard tensile test is conducted to failure on specimens that survive this preconditioning. The parameter measured is the percentage area of adhesive failure.

Some initial results indicating the relative influence of each parameter on the level of adhesive failure are shown in Figure 12.

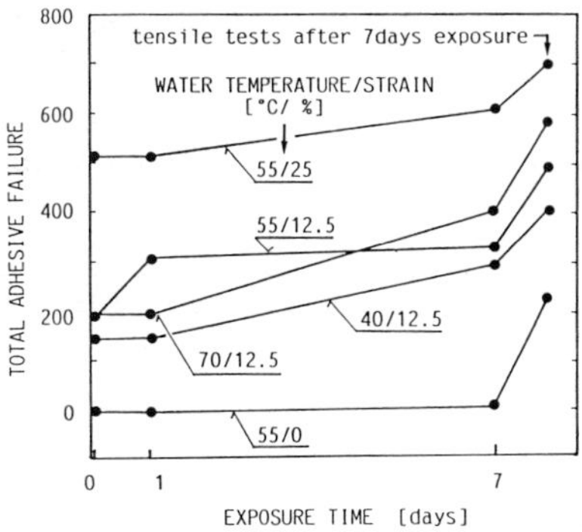


FIG. 12--Plot of total percentage adhesive failure (% TAF) for 13 sealant/substrate pairs versus immersion time for five different water temperature/fixed strain combinations [15].

These data indicate that the imposition of a load during the exposure, in this case in the form of a fixed strain, increases the severity of the test, with adhesive failures occurring during the first hour of water immersion. In each case, tensile tests on surviving specimens result in a higher level of adhesive failure than the 7 days immersion without strain. Generally the level of adhesive failure increases with increasing water temperature, strain or exposure time. The results also show that the level of strain has a greater relative effect than temperature. Furthermore, when bond failure occurs during the immersion period, it does so very early in the test and this provides the potential for a much quicker procedure than those being considered elsewhere.

Life Cycle Prediction

The ultimate objective is to develop a test protocol which can be used as the basis for a performance specification and life prediction. By varying the severity of the conditions, it should be possible to determine the relative level of performance rather than whether a particular combination passes or fails. In this way candidate materials can be ranked and correlations explored against field performance. While it is far from a procedure for life cycle prediction, the investigations are providing an improved basis for selection against performance parameters.

**SEALANT PERFORMANCE MONITORING**

Despite the successful record of Silicone Structural Glazing (SSG) technology, which was first introduced in 1968 in USA, there are noticeable gaps in knowledge regarding long-term durability of these materials. As a result of this, extensive monitoring programs are employed in order to gather information on performance of silicone sealants and to assure the structural soundness of existing facades erected with the use of SSG.

A monitoring program was first (in Australia) set up for 222 Exhibition Street in Melbourne by Ove Arup and Partners and Cupples Products. Both CSIRO and Arup Facade Engineering have contributed to this project's monitoring.

The Council of the City of Sydney were the first to provide definite rules on monitoring. Approximately 3 years ago the Council involved a group of experts experienced in structural glazing eg. Arup Facade Engineering, Peter Hartog, Connell Wagner, BOMA, sealant manufacturers and others to form a committee. A set of guidelines, which covered both existing curtain walls and those to be built has been developed. The guidelines comprise the following:

(a) Confirming initial adequacy of existing curtain walls. The council guidelines classify a curtain wall according to the documentary evidence of its quality, as follows:

A: full quality documentation for design and installation,
B: minimal documentation, eg. fabrication drawings and sealant manufacturer's records,
C: no documentation. This, unfortunately, is an all-to-common situation.

The building's classification determines the level of monitoring. Monitoring consists of periodic visual inspections, material sampling and de-glazing.

Inspections carried out by Arup Facade Engineering have occassionally revealed the occurrence of the following typical problems:

(i) joints with excessive voids; "aero bar" void; (ii) sealant which is brittle and breaks easily, rather than having the usual extensibility and recovery of good quality silicone; (iii) vision panels without structural sealant; the glass being retained by double-sided tape and a silicone weather seal.

So far no evidence of structural sealant deterioration was found.

One of the problems encountered is to determine the quality of the silicone joints of buildings in category C. For some of these, no documentation exists regarding the type of silicone.

Structural engineers are constantly at odds with the chemical engineers from the sealant companies as to how this is to be done. They don't like the idea of testing sealant to destruction, because this implies it can break. Also, they claim rightly, that a good sealant joint will fail at a safety factor of 4.5 to 5.5 times design (tension) load. They are then quite comfortable in saying "despite what some structural engineers say, if a 1.5 times design load is applied to a structural silicone glazed installation, and it does not fail, then this proves it is adequate." However, the non-destructive testing at the load equal the said 1.5 times design load does not prove that the real safety factor is 5 and not for instance 1.85!

(b) Experimental techniques for determining actual (in-situ) mechanical properties and performance characteristics of sealants.

Arup Facade Engineering have developed a simple technique for determining actual strength values for in-situ seals. This involves:

(i) Deglazing a panel, leaving half the sealant on the panel and half on the frame; (ii) Use a thin layer of 2-part silicone to bond short 'T' sections to remaining silicone, and re-make the sealant joint back to original shape; (iii) Use a small portable tensile testing rig, eg. with a load cell and electronic readout to assess strength; (iv) Test sealant on both substrates using 3 to 6 samples of each.

We believe this to provide a reasonable comparison of seals, however we do not yet regard it as an absolute sealant strength measurement. One reason for this is that the extension is applied by hand and is neither even nor consistent. Also, we have not yet researched the effect of the two-part sealant on the existing structural sealant under assessment. Most of the information gained in this exercise is revealed in the deglaze. The test does, however give a good indication of the most likely adhesive or cohesive failure mode.

**REFERENCES**

[1] LALAS, P., "Engineering Curtain Walls:Structural Silicone", in The Performance of Sealants and Adhesives in Building Facades, School of Architecture, UNSW, 21 Nov. 1989.

[2] TIMOSHENKO, S., WOINOWSKY-KRIEGERS, "Theory of Plates and Shells", 2nd Edition, McGraw Hill, Chapter 5 Part 27, 1959.

[3] MINOR, J.E., "Developments in the Design of Architectural Systems". Proceedings of the First Structural Engineering Conference, IEAus, pp.77-81, 1987.

[4] LALAS, P., Confidential Report. Arup Facade Engineering, Sydney, N.S.W.

[5] SANDBERG, L.B., AHLBORN, T.M., "Combined Stress Behavior of Structural Glazing Joints", Journal of Structural Engineering, Vol. 115, No. 5, pp. 1212-1224, May 1989.

[6] GUTOWSKI, W.S., CSIRO Report-DBCE Doc 92/100(M)-"Structural Glazing Project:Cohesive Properties of Silicone Sealants and Self-Adhesive Tapes", Melbourne, p. 20, 1992.

[7] Test Methods for Determining Tensile-Adhesion Properties of Structural Silicone Sealants. ASTM C1135 Standard.

[8] KARPATI, K.K., "Mechanical Properties of Sealants III:Peformance Testing of Silicone Sealants", Journal of Paint Technology, Vol. 44, No. 571, August 1972, pp. 75-85.

[9] FERRY, J.D., "Viscoelastic Properties of Polymers", John Wiley & Sons Inc., New York, London, Sydney, Toronto, 1970.

[10] SMITH, T.L., "Viscoelastic Behaviour of Poliisobutylene under Constant Rates of Elongation", Journal of Polymer Science, Vol. XX(1956), pp. 89-100.

[11] KLOSOWSKI, J.M., "Sealants in Construction", New York, Basel: Marcel Dekker, Inc., p. 194, 1989.

[12] IKER, J. WOLF, A.T. "Comparison of European and US Testing/Qualification Procedures for Structural Glazing Silicone Sealants". Symposium on Building Sealants: Materials Properties and Performance, ASTM STP 1069, Thomas F. O'Connor, ed., Philadelphia, ASTM, pp. 67-78, 1990.

[13] Standard Test Method for Adhesion-in-Peel of Elastomeric Joint Sealants, ASTM C794-80 Standard.

[14] CERRA, A., GUTOWSKI, W.S., "Performance-Based Testing of Structural Sealants", Science and Technology of Building Seals and Sealants: Fifth Volume, ASTM STP 1271, Michael Lacasse, Ed., American Society for Testing and Materials, Philadelphia, 1995.

[15] CERRA, A., "Development of New Accelerated Adhesion Test for Sealants", Journal of Testing and Evaluation. Accepted for publication in 1995.

Richard Wm. Tock,[1] and Ramesh Keshavaraj [2]

# PHYSICAL-CHEMICAL CHANGES IN SILICONE ELASTOMERS USED FOR BUILDING SEALANTS

---

**REFERENCE:** Tock, R. Wm., and Keshavaraj, R., **"Physical-Chemical Changes in Silicone Elastomers Used for Building Sealants,"** Science and Technology of Building Seals, Sealants, Glazing, and Weatherproofing: Fifth Volume, ASTM STP 1271, Michael A. Lacasse, Ed., American Society for Testing and Materials, 1996.

**ABSTRACT:** The modulus of a silicone sealant elastomer is one of the most fundamental properties of these materials that can be measured on a routine basis. It is also a very complex function of several variables. These variables include the chemistry of the silicone polymer and the type of filler compounded into the product. Moreover, once the sealant is placed into service, additional variables such as loading and displacement history, time-temperature cycles, and weathering exposure become significant. Several approaches have been developed to monitor and characterize changes observed in three specific commercial silicone sealants. While silicone polymerization chemistry determines the broad characteristics of these sealants, it was observed that filler type, particulate bulk and cohesive material performance, and weathering can also contribute long term effects. These effects on some of the engineering properties pertaining to silicone elastomers used in building sealant application is the subject of this paper. Elastomers exhibit compressive moduli that can be several times that of the tensile moduli. In addition, acid and base pH levels from exposure to precipitation can accelerate residual cross-linking and ultimately lead to sealant degradation. The net result produced by such experiments has been a better understanding of sealant properties.

**KEYWORDS:** structural silicone sealants, modulus, fillers, weathering, aging, tensile strength

---

[1]Professor, Department of Chemical Engineering, Texas Tech University, Lubbock, TX 79409-3121.

[2]Research Associate, Department of Chemical Engineering, Texas Tech University, Lubbock, TX 79409-3121.

## INTRODUCTION

This paper represents a continuation of earlier experimental work conducted on the aging of structural silicone sealants due to the effects of environmental exposures [1]. Basically, the earlier work was concerned with the predictability of cohesive failure in three different commercial, structural silicone sealants following their exposure to selected environmental conditions. Nominally such sealants when used in the United States are expected to have a lifetime of fifty years, while applications in the EEC are required to perform for a least one hundred years. These spans represent significant time periods. Hence, it had been suggested that the sealants might not exhibit this level of longevity if they were repeatedly exposed to the increased acid rain or elevated ozone concentrations typical of today's atmosphere. For this reason, and because of the rapid development and extensive use of these sealants in modern high rise buildings, some evaluation of the potential risks associated with the premature failure of the sealants was desirable. In a combine effort between Dow Corning, NSF, and Texas Tech University, an attempt was begun in the late 1980's to quantify climatic factors which could contribute to the risk of premature failure for structural silicone sealants. This study has been the source of several published papers [2-6]. For the study presented in this paper, many small samples and several full scale IG glazing units were assembled in Midland, Michigan, and tested at Texas Tech University's Glass Research Institute. This project and other similar studies at other laboratories during the 1990's were part of the ASTM sponsored symposia on building sealants [7]. The net result from these studies has been that we now have a better understanding of sealants properties and performance, and, perhaps more importantly, the sealant manufacturers have continued to provide architects and engineers with better products for structural applications.

## RESEARCH EFFORTS

The research findings reported in this paper deal with changes in the modulus or stiffness of three commercially available silicone sealants: DC-790, DC-995, and DC-983. Results from earlier investigations on the aging characteristics of these three sealants can be found elsewhere [8-12]. Some new tensile strength data will be discussed for these three sealants, as will some thoughts about the fillers used in current silicone sealant formulations. Moreover, as indicated earlier, the data presented in this paper pertain to bulk, cohesive material performance. Intentionally excluded were the interfacial adhesive properties of the tested sealants.

Based on data acquired with the full scale IG units tested by the Texas Tech University Glass Research Institute, it appeared that commercial sealants which exhibit the largest moduli (greatest stiffness) give the best performance in structural glazing and IG applications [2,3]. Justification for this declaration followed this pattern of thought: glass is a stiff, surface crack sensitive material which exhibits small strains at failure. These characteristics could lead to a scenario wherein the large glass plates used in glazing would possibly deform. Failure under high wind loads would be enhanced if the stiffness of the sealant was insufficient to help resist the loading. This possible scenario lead to the following questions: (1) How stiff are commercial sealants, (2) How does this stiffness change with time, loading and environmental exposure, and (3) To what extent is the stiffness predictable from known material properties.

In order to answer these and other questions a systematic approach utilizing both theoretical and empirical models was developed [9]. For example, the silicone sealants were first assumed to behave as ideal elastomers even though they are composite materials [3]. Time dependent behavior over relatively short time periods was characterized with a viscoelastic Maxwell model in order to visualize stress relaxation behavior [8]. A series chemical reaction model was used to explore changes in chemical crosslink density with time due to aging [11,14]. Numerous experimental tensile and compression data were taken on sample coupons which had been exposed to weathering [8]. In this paper, results are presented on tensile tests to rupture for weathered samples in order to determine failure responses to weathering exposure.

## EXPERIMENTAL APPROACH

### Theory

For an ideal elastomer the stress in tension or compression is given by the following expression:

$$\sigma = RTN\left[\alpha - \frac{1}{\alpha^2}\right] \tag{1}$$

This equation includes the independent variables of absolute temperature (T), a measure of the crosslink density (N), and the extension ratio ($\alpha = l/l_o$). The modulus for the same idealized elastomer can be obtained by differentiating stress in the preceding expression with respect to $\alpha$.

$$E = \frac{d\sigma}{d\alpha} = RTN\left[1 + \frac{2}{\alpha^2}\right] \tag{2}$$

Equation (2) indicates that, like stress, the stiffness of the idealized elastomer is also a function of absolute temperature, crosslinking, and extension ratio. Of these three variables, crosslink density and extension ratio have the most significant influence on the modulus of products used in the majority of the glazing applications. This is because excursions in temperature from 25°C (298°K) are not likely to exceed 20% on the absolute temperature scale. Hence, modulus can be expected to change by only plus or minus 20% due to temperature. However, it will be shown that changes in crosslinking and/or strain can cause the modulus to increase or decrease by as much as two to three times, particularly during compression loading.

What these idealized expressions (Equations 1 and 2) do not include, however, are the effects on the sealant's modulus due to the use of fillers. For reasonably low levels of filler loading and homogeneous dispersion (10-15% by Vol), it is possible to assume that the particulate fillers behave as small dispersed sites of highly crosslinked elastomer. If this were the case, Equations (1) and (2) would retain some validity, which they do as will be discussed later. However, it is easier to use other expressions, such as the Kerner equation [15], to predict filler effects on the modulus of composites produced by compounding particulate materials into a polymer matrix. This was what was done in this study. An expression for a modulus derived by Smallwood [15] for small amounts of filler (<15%) is as follows:

$$E = \frac{\mu(3\lambda + 2\mu)}{(\lambda + \mu)} \tag{3}$$

Equation 3 was derived from the equilibrium equations in terms of the bulk modulus, $\lambda$, and shear modulus, $\mu$, of an elastomer. Using the subscript p for the matrix (i.e. polymer) and f to denote the particulate filler, the Kerner equation for the bulk and shear modulus can be written as follows [17]:

$$\lambda = \frac{\left[\dfrac{\lambda_f v_f}{3\lambda_f + 4\mu_p}\right]}{\left[\dfrac{v_f}{3\lambda_f + 4\mu_p}\right]} \tag{4}$$

and

$$\mu = \mu_p \left\{ \frac{\left[\dfrac{\mu_f v_f}{(7-5v_p)+(8-10v_p)\mu_f}\right] + \dfrac{v_p}{15(1-v_p)}}{\left[\dfrac{\mu_p v_f}{(7-5v_p)+(8-10v_p)\mu_f}\right] + \dfrac{v_p}{15(1-v_p)}} \right\} \tag{5}$$

In the above equation, $v_f$ is the volume fraction of the filler and $v_p$ is the Poisson's ratio for the polymer, $\lambda_p$ and $\mu_p$ refer to the polymer, $\lambda_f$ and $\mu_f$ refer to filler component. It should be remembered that Equation 5 excludes the matrix phase. Equations 4 and 5 can be reduced to much simpler expressions by using an ideal elastomer assumption. Young's modulus can be calculated by substituting Equations 4 and 5 in Equation 3.

Yield Strength and Elongation

The strength and elongation of filled silicone elastomers are much more difficult to quantify. Typically, breaking strength is a statistical property. Moreover, most elastomers yield before they ever reach failure stress levels. In the classical case for low volume fractions of filler where good interfacial adhesion exists, the yield stress and the elongation at rupture of the composite sealant are equivalent to the neat silicone rubber [15]. However, as the volume fraction of the particulate compounded into the silicone matrix increases and approaches a maximum, (approximately 0.60 for small randomly packed particles), both failure stress and elongation rapidly diminish and approach zero. Sealants are not compounded with more than 15% by volume of filler, however. This creates a paradox for particulate composites; increases in particulate loading can enhance stiffness, but particulate fillers nearly always diminish strength and elongation. This behavior is reflected in the following approximate equations for elongation and yield strength [16].

$$\varepsilon_c = \varepsilon_m \left[ 1 - \left( \frac{V_f}{P_f} \right)^{\frac{1}{3}} \right] \tag{6}$$

or

$$\sigma_{yc} = \left( 1 - \frac{V_f}{P_f} \right) \sigma_{ym} \cdot A \tag{7}$$

In Equations (6) and (7), the subscripts c and m refer to composite and matrix materials, while the volume fraction, $V_f$, and packing factors, $P_f$, are related in the following manner.

$$0 \leq V_f \leq P_f = 0.60 \tag{8}$$

Also, A, in Equation (7) represents an adhesion factor which varies from 0.8 to 1.1 with no adhesion, and never exceeds 2.0 even with the use of good adhesion promoters. Hence, yield and breaking stresses are diminishing properties as volume loading of filler increases.

## Experimental Program

Test coupons were prepared as per the ASTM standards from cast sheets of the three commercially available silicone rubber sealants: DC-790, DC-995, and DC-983, listed in the order of increasing stiffness. Sixty samples were cut from each material and were subjected to artificial weathering; namely moisture at four different pH levels (3 to 11), sunlight and ozone. Each set of test coupons were exposed to moisture at a particular pH level and then were allowed to dry in the outdoors on a rack. This sequence of spraying followed by drying outdoors was refered to as one weathering event in our earlier papers [8-12], and the same terminology will be followed here. According to our test procedure, the test coupons were removed from the experimental layout at regular recorded intervals and were tested at room temperature for changes in their properties. The samples were then reinstalled in the experimental layout for further exposure. It was assumed and later verified through experimentation that the short duration tests did not cause any change in the crosslinking mechanism. For the three types of sealants, each sample was tested with five sub-sample tests and three sample replications in each case. Several coupons were also kept at ambient laboratory conditions as a control. More details about this experimental program can be found elsewhere [8-12].

## RESULTS AND DISCUSSION

### Filler Effects

Fillers have the capability to significantly increase the stiffness or modulus of the silicone sealants. Three different fillers are used with silicone elastomers. The fillers are: (A) crushed calcium carbonate, (B) crushed amorphous silica, and (C) carbon black. A summary of the properties of different fillers is given in Table 1.

Table 1 -- Some properties of fillers used in silicone rubber sealants

| Type of Filler | Hardness (Mohs) | Environmental Chemical Resistance |
|---|---|---|
| Crushed calcium carbonate | 3 | weak in acid pH |
| Amorphous silica powder | 7 | weak in basic pH |
| Crystalline silica powder | 7 | resistant to acid & basic pH |
| Carbon blacks | <2 | UV absorber |

Only the carbonate and the silica are used to stiffen the matrix elastomer. Carbon black is used as UV inhibitor. Of the two mineral fillers, powdered silica has the highest hardness, (7 Mohs versus 3 Mohs for the carbonate). For this reason the silica will provide more stiffness at the same fraction of volume loading. Silica may also bond better than the calcium carbonate because of the silane chemistry of the silicon elastomer. However, since both fillers have elastic moduli that are three orders of magnitude greater than the neat silicone matrix, both are frequently used to stiffen the sealant elastomers. Figure 1 shows this hypothetical stiffening effect as a function of the fractional volume loading of the filler. The increase in modulus is linear up to a volume faction of 30%. At this level the composite's modulus is five time that of the unfilled polymer. For compounded products with volume fractions in excess of 30%, the modulus increases even more rapidly. It approaches a modulus that is in excess of 20X that of the silicone elastomer as the volume loading of the filler nears the fractional, 0.60, theoretical limit. But it should be remembered that in the case of sealants, compounding approaches a maximum of 15% by volume and, hence, the increase in stiffness is less than 4X. Also, the increase in stiffness is linear with compounding in this range.

Since both fillers are effective modifiers of the modulus, selection of one filler over the

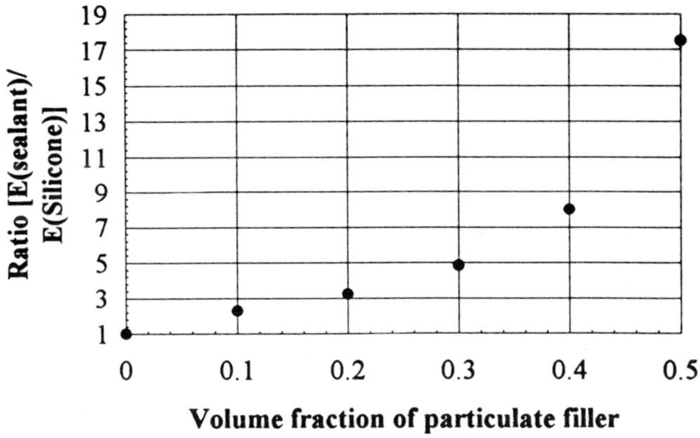


FIG. 1--The effect on modulus caused by increased filler fraction

other is based on cost and/or their respective environmental chemical resistance. The calcium carbonate is cheaper, but it is more reactive with the deleterious effects encountered in acid rain environments. The silica, if it is in the crystalline form, is highly resistant to both acid and basic environments. However, amorphous silica is attacked by basic solutions like those used for window washing fluids, or those produced naturally by rain in arid regions where alkaline soils predominate. Hence, for longevity, the filler should be tailored to the expected environment.

Crosslinking Effects

As was mentioned earlier, it is possible for the degree of crosslinking to significantly effect the modulus and strength of elastomers. While this fact is well known and publicized for natural and synthetic elastomers, it is usually not given much consideration with silicone sealants. The structural seals are usually placed into service and then forgotten about. By using Tension-Adhesion (TA) joint specimens in this study, we were able to expose the sealants materials to different weathering environments and then test for changes in modulus and/or crosslink density. It was something of a surprise to find that both high and low pH mixture caused significant increases in modulus [8]. These increases were attributed to continued crosslinking of residual sites catalyzed by the acid and/or basic pH levels. Figures 2 and 3 show these effects for simulated weathering. During these tests, moisture, with adjusted pH levels, was daily spayed on the samples which were then allowed to dry in air. The modulus and degree of crosslinking for each of the three sealant materials was shown to rise and then fall over time. Typically the maximum modulus reached was 4X the initial base modulus. No additional weathering moisture was applied

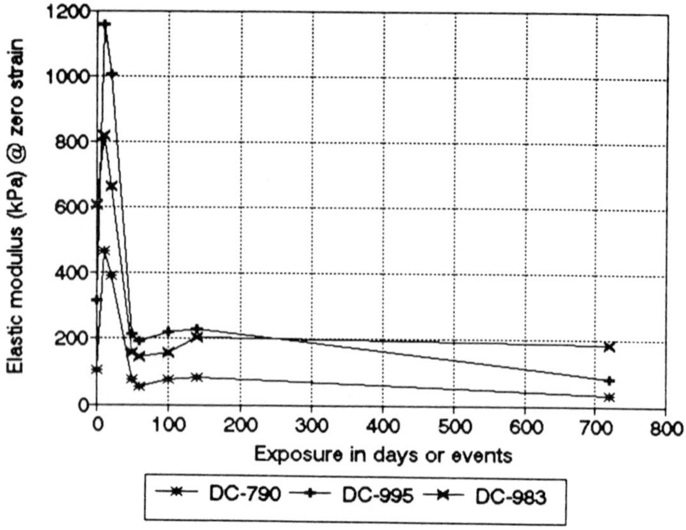


FIG. 2--Weathering effects on the elastic modulus of sealants

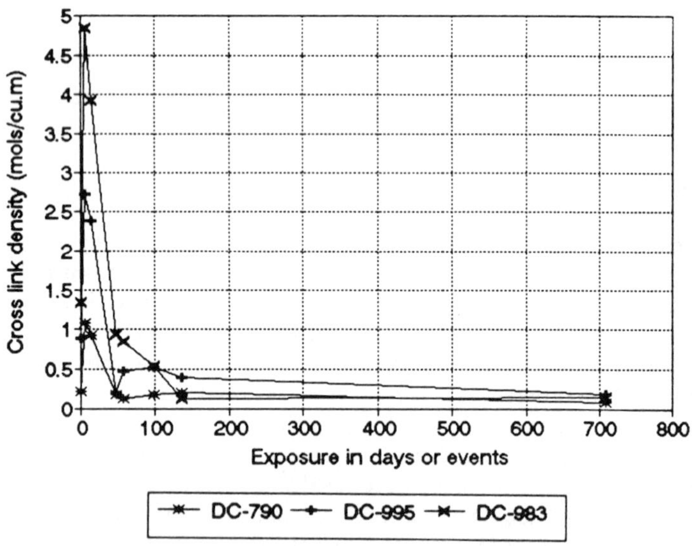


FIG. 3--Change in crosslink density caused by weathering exposure

after an initial period of 60 days or sixty applications (events). The data points which were connected by computer generated lines indicated that the silicone sealant samples continued to decline slightly over time in both their modulus and degree of crosslinking. This was not the case with tensile strength, however.

As mentioned earlier, a chemical-kinetic model was developed for the changes in the crosslink density due to weathering. These changes can be easily correlated to the changes in the stiffness of the sealant. An assessment of the progressive level of crosslinking based on a series of hypothesized chemical reactions [9,14] is shown in Table 2 for the three sealants for comparison purpose.

Table 2 -- A Theoretical assessment of the level of crosslinking observed for the three silicone sealant samples.

| Sample Designation | General Description | Total Sites Available | Initial % Crosslinked | Maximum % Crosslinked |
|---|---|---|---|---|
| DC-790 | Low Modulus | 1.2 mol/m$^3$ | 15 | 87 |
| DC-995 | Medium Modulus | 3.3 mol/m$^3$ | 21 | 73 |
| DC-983 | High Modulus | 6.3 mol/m$^3$ | 18 | 72 |

The possible crosslinking in service depends on the number of uncrosslinked sites available in a particular sealant before installation. In general, the low modulus sealant exhibited a much higher increase in the crosslinked sites with time in relation to either the medium or high modulus products. Hence, a very low modulus sealant is not used nor does Dow Corning suggest such type of products in glazing applications. However, a knowledge of the total number of sites available for crosslinking in a sealant before installation is very important irrespective of the type of sealant used.

Tension versus Compression Loading

Unlike most structural materials, elastomers exhibit a significantly different modulus when placed in compression as opposed to tension loading. This behavior of the modulus as a function of strain is shown in Figure 4.

The ratio of the modulus to the modulus exhibited at zero strain was calculated using Equation (2). As shown, a significant increase in modulus is experienced with relatively small compressive strains, while the modulus decreases with tensile loads. For glazing, compressive loads are usually due to wind loading. The effects of initial loading on the oxidative effects of sealants was the subject of our earlier publication [12]. In this reference, the synergistic effect of loading and oxidation on sealants are discussed.

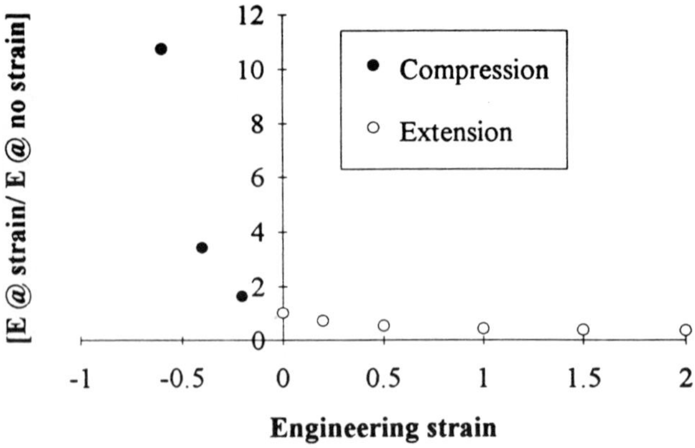


FIG. 4-- The effect on sealant modulus due to loading orientation

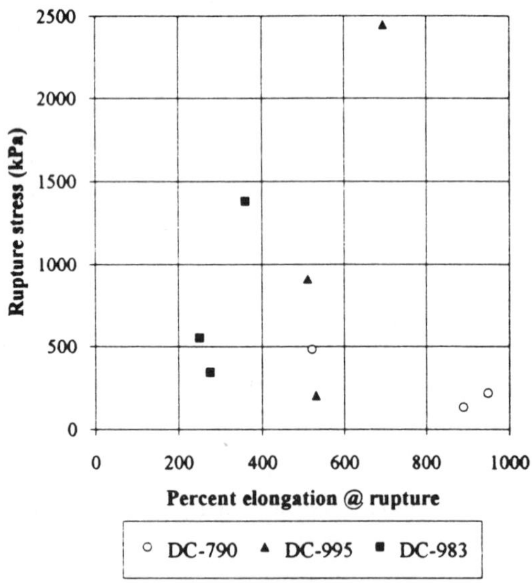


FIG. 5--The effect of weathering on failure tensile stress

Ultimate Failure Stress

The ultimate tensile failure stress as a function of ultimate elongation is depicted in Figure 5 for the samples exposed to artificial weathering. These data are displayed as only three points which represent three different periods of weathering exposure. The lowest rupture stress for each type of sealant was observed for the "as received" sealant materials. Following exposure to artificial weathering for sixty days, all three sealants experienced an increase in breaking strength. Only the low modulus material, DC-790, displayed a subsequent lost in tensile elongation (but not strength) after 720 days. With added exposure time, the medium and high modulus silicone sealants are expected to show a pattern similar to that of the low modulus sealant. This behavior is typical of elastomers which degrade [17].

## CONCLUSIONS

From the experimental data generated in this study, it has been shown that certain commercial silicone sealants can expect to undergo significant changes in modulus with time in service. The observed increases in modulus due to crosslinking suggests the need for further investigation to determine the significance of crosslink changes density in actual use. However, the simulated weathering tests indicated that prolonged exposure to adverse environments can hasten property deterioration. It is not clear, however, if this behavior will compromise performance assurances. It may be necessary for sealant materials to be shielded from direct environmental contact, in order for them to meet such expectations.

## ACKNOWLEDGMENTS

Financial support from the U.S. National Science Foundation (Grant MSS 900 7911) and Dow Corning Corporation's Elastomers Research Laboratory is gratefully acknowledged.

## REFERENCES

[1] Keshavaraj, R., and Tock, R. Wm., "Engineering Property Changes in Silicone Sealants Due to Aging," SPE. Conf. ANTEC'93- Technical Papers, Vol. 3, 2906-2910, 1993.

[2] Vallabhan, C.V.G., et al., "Thin Glass Plates on Elastic Supports", Journal of Structural Engineering, ASCE 3(11) 24-2426, 1985.

[3] Tock, R. Wm., et al., "Viscoelastic Properties of Structural Silicone Rubber Sealants," Advanced in Polymer Technology, 60 May/April, 78-88, 1987.

[4] Bailey, J.R., Minor, J. E., Tock, R. Wm., " Changes in Seal Shapes of Structurally Glazed Insulating Glass Units," in Building Sealants: Materials, properties, and

Performance, STP 1069, ASTM, Philadelphia, PA, pp 42, 1990.

[5] Vallabhan, C. V. G., and Chou, G. D., "Sealant Stress in Structural Glazing: A Mathematical Model," in Building Sealants: Materials, properties, and Performance, STP 1069, ASTM, Philadelphia, PA, pp 22, 1990.

[6] Tock, R. Wm., "Temperature and Moisture Effects on Engineering Properties of Silicone Sealants," in Building Sealants: Materials, properties, and Performance, STP 1069, ASTM, Philadelphia, PA, pp 167, 1990.

[7] O'Conner, T.F., Editor, Building Sealants: Materials, properties, and Performance, STP 1069, ASTM, Philadelphia, PA, 1990.

[8] Keshavaraj, R., Tock, R. Wm., and Vallabhan, C. V. G., Effects of Moisture on Structural Silicone Rubber Sealants Used in Window Glazing Applications," Construction and Building Materials, 8(4), 227-232, 1994

[9] Keshavaraj, R., and Tock, R. Wm., "Modeling of Crosslinking Mechanisms When Structural Silicone Sealants and Subjected to Moisture," Polymer-Plastics Technology and Engineering, 32(6), 579-593, 1993.

[10] Keshavaraj, R., and Tock, R. Wm., "The Effect of Length in the Test Section on the Mechanical Performance of Silicone Sealants Used in Structural Glazing," Journal of Materials Science Letters 12, 1627-1628, 1993.

[11] Keshavaraj, R., and Tock, R. Wm., "Changes in Crosslink Density of Structural Silicone Sealants Due to Ozone and Moisture," Polymer-Plastics Technology and Engineering, 33(4), 397-417, 1994.

[12] Keshavaraj, R., and Tock, R. Wm., "Oxidative Effects of Ozone on the Aging of Structural Silicone Elastomers," Advances in Polymer Technology, Vol. 13, No. 2 149-156, 1994.

[13] Nagappan, N., Behavior of Laterally Loaded Structurally Glazed Systems, MS Thesis, Texas Tech University, Lubbock TX 79409-1023, 1992.

[14] Keshavaraj, R., and Tock, R. Wm., "Modeling of Changes in Crosslinking for Structural Silicone Sealants Subjected to Moisture and Sunlight," Polymer-Plastics Technology and Engineering, 33(5), 537-550, 1994.

[15] Katz, H.S., and Milewski, J.V., Handbook of Filler for Plastics 34, Van Nostrand Reinhold, New York, 1987.

[16] Katz, H.S., and Milewski, J.V., op.cit., 38-40.

[17] Treloar, L.R.G., The Physics of Rubber Elasticity, 3rd ed., Clarendon, Oxford, England, 1975.

K. Frank Chang[1] and L. Bogue Sandberg[2]

# STRUCTURAL BEHAVIOR OF SILICONE BONDED GLASS BLOCK PANELS

**REFERENCE:** Chang, K. F. and Sandberg, L. B., **"Structural Behavior of Silicone Bonded Glass Block Panels,"** Science and Technology of Building Seals, Sealants, Glazing, and Waterproofing: Fifth Volume, ASTM STP 1271, Michael A. Lacasse, Ed., American Society for Testing and Materials, 1996.

ABSTRACT: Silicone sealant was substituted for mortar in bonding glass blocks. The sealant's tensile and shear strengths and stiffnesses were determined. Joints bonding two glass blocks were tested for stiffness and strength in tension, bending, out-of-plane shear, and in-plane shear. Bending tests were done on specimens one block wide and four blocks long to evaluate one-way bending behavior. A six block by six block panel, supported on all four sides, was built and tested under simulated wind load. An analytical model with material nonlinearity in the joints was developed for the one-way bending case. It gave good comparisons with the experimental data to load levels approaching failure. A more complex analytical model was developed for the two-way panel. It was only valid for lower load levels, in the range of potential allowable design loads, but compared well with test results. Silicone bonded glass block panels have potential for meeting the wind load requirements necessary for exterior use.

**KEY WORDS:** Glass, Block, Silicone, Sealant, Structural Glazing

Traditionally, glass block wall panels been constructed using masonry mortar in the joints between the block. There is at least one proprietary system in which the

[1]Structural engineer with Structural Engineering Associates, Inc., 3838 N.W. Loop 410, Suite 200, San Antonio, TX 78229

[2]Professor of Civil and Environmental Engineering, Michigan Technological University, Houghton, Mich.

blocks are placed in an aluminum grid, with a silicone sealant used to seal the joints between the faces of the blocks. This paper presents a study of the use of silicone sealant as the sole replacement for conventional mortar. In such a system the silicone sealant carries all loads between blocks as well as providing a weather seal.

## EXPERIMENTAL INVESTIGATIONS

In the study on which this paper was based [1], two sealants were evaluated, Dow Corning DC 999 and Dow Corning Glass Block Sealant. Behavior was similar for the two sealants. Since the Glass Block Sealant was slightly stiffer and stronger, it will be the focus of the present paper.

### Sealant Material Properties

The first step was to characterize the basic stiffness and strength characteristics of the silicone sealant. This was done by testing five joint specimens in tension and five in shear. These specimens were cured based on a schedule suggested by the sealant manufacturer. They cured 4 weeks at an average temperature of 21° C (70° F) with relative humidity averaging 47 percent. Then they were post-cured for an additional three weeks at 47° C (117° F) in a small oven heated with two thermostatically controlled 100w light bulbs. A pan of water with a wick above it provided moisture in the oven. This was done to accelerate the slow residual cure typical of the sealants. Aluminum substrates were used. The bite to width ratio on these specimens was 6.35 mm by 6.35 mm (0.25 in by 0.25 in) with a length of 200 mm (8.0 in), approximating the actual glass block joints. Substrates were rigidly attached to the testing machine cross heads so only a pure extension or shear racking mode of displacement occurred. The testing was done at 1.3 mm/min (0.05 in/min), 21° C (70° F), and approximately 47% RH. A complete record of these and all subsequent test data are given in [1]. The data was very consistent and all failures were cohesive. The curves shown in Fig. 1 and Table 1 were generated from the data by trial and error regression on a linear transformation [2].

### Tests on Single Joint Specimens

The second set of experiments focused on the stiffness and strength of a single joint (with front and back beads) between two glass blocks. Specimens consisted of two blocks, a clear vinyl spacer, and the sealant beads on each face (Fig. 2). The blocks, furnished by Pittsburgh Corning Co., were 197 mm on a side and 99 mm in overall width (nominal 8 in x 8 in x 4 in). The surfaces to be bonded were cleaned with isopropyl alcohol to remove an edge coating applied during manufacture. The spacer was cut to a 197 mm (7.75 in) length. It served to set the sealant joint width and also set the bite, by acting as a backer rod. Otherwise, it appeared to have little or no effect on structural performance. The resulting beads had an average bite of 5.1 mm (0.20 in) and a width of 4.1 mm (0.16 in). These and all subsequent specimens were cured, for a minimum of 8 weeks, and then tested at ambient

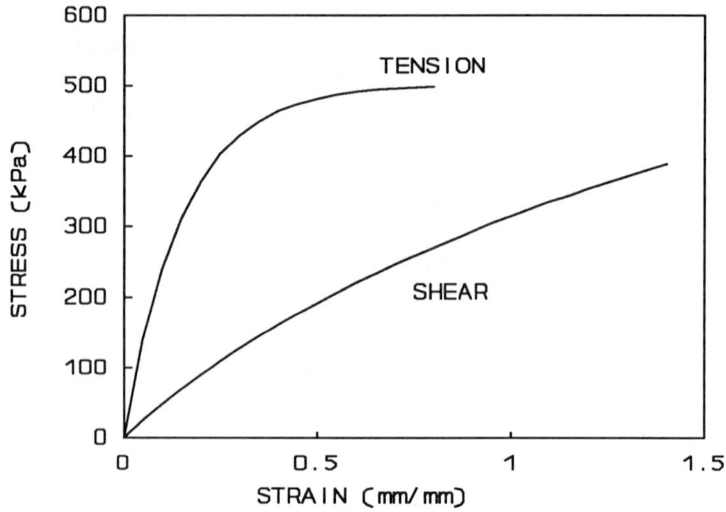


FIG. 1—Stress-strain curves from 6.4 mm by 6.4 mm joints.

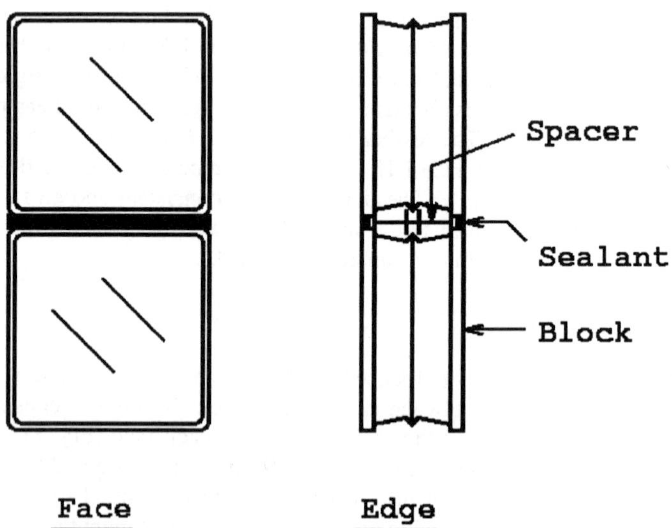


FIG. 2—Single joint specimen schematic showing joint detail.

conditions averaging 21° C (70° F) and 47% RH.

Single joint specimens were tested in tension, bending, transverse shear, and longitudinal shear. The test speed was 1.3 mm/min (0.05 in/min) There were four specimens for each type of test. The tests are shown schematically in Fig. 3. In these tests and all that follow, bond of the sealant to the glass was excellent with little or no adhesion failure. The test results were fitted with the exponential curves listed in Table 1 along with their coefficients of determination. The curves are plotted in Fig. 4.

A separate set of single joint specimens were made with a modified spacer. The spacer edges each had two rectangular cutouts to increase the silicone bond area by 50 percent. The resulting strength increases were 47%, for tension, 31% for bending, 41% for transverse shear, and 50% in longitudinal shear. The bending and transverse shear strength are the most important. The fact that the strength increases are less than the increase in bond area undoubtedly stems from the stress concentrations induced by the notched geometry.

TABLE 1--Regression Analysis for Material and Single Joint Data

| Test | Fitted Equation | $R^2$ |
|---|---|---|
| Material -- | | |
| Tension, kPa: | $\sigma = 501\ (1 - \exp(-6.5\varepsilon))$ | 0.96 |
| Shear, kPa: | $\tau = 346\ \ln(1 + 1.48\gamma)$ | 0.99 |
| Single Joint -- | | |
| Tension, kN: | $T = 0.971\ (1 - \exp(-1.48\Delta_T))$ | 0.96 |
| Bending, N-m: | $M = 71.5\ (1 - \exp(-59.1\tan(\theta/2)))$ | 0.88 |
| Trans. Shear, kN: | $V_T = 0.971\ (1 - \exp(-0.230\Delta_{TS}))$ | 0.98 |
| Long. Shear, kN: | $V_L = 0.617\ (1 - \exp(-0.807\Delta_{LS}))$ | 0.94 |

$\varepsilon$ = tensile strain, $\gamma$ = shear strain, $\Delta$'s = displacement in direction of load and $\theta$ = joint rotation angle

## One-way Bending Tests

To simulate the wind load behavior of a long, narrow glass block panel, sets of four

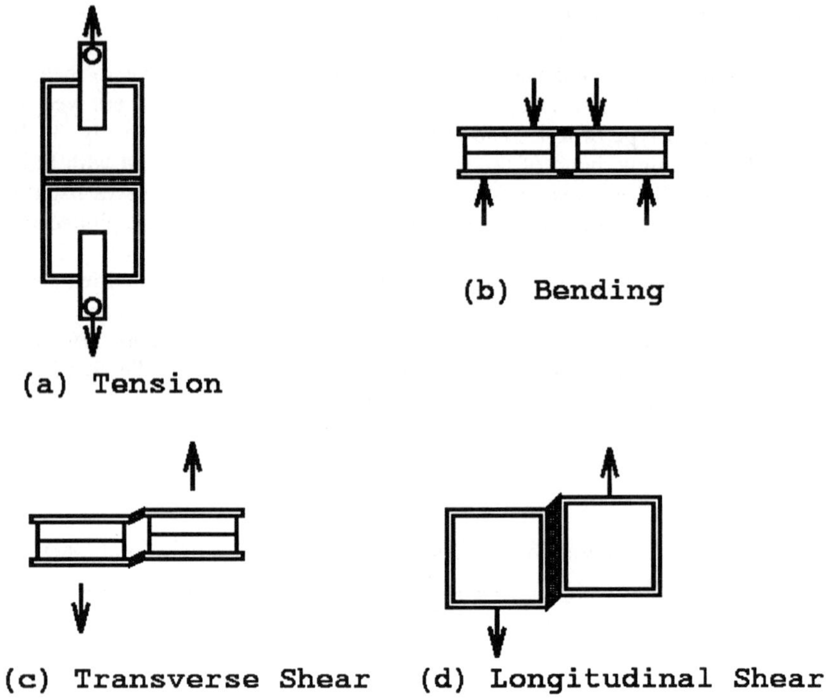


FIG. 3—Single joint tests.

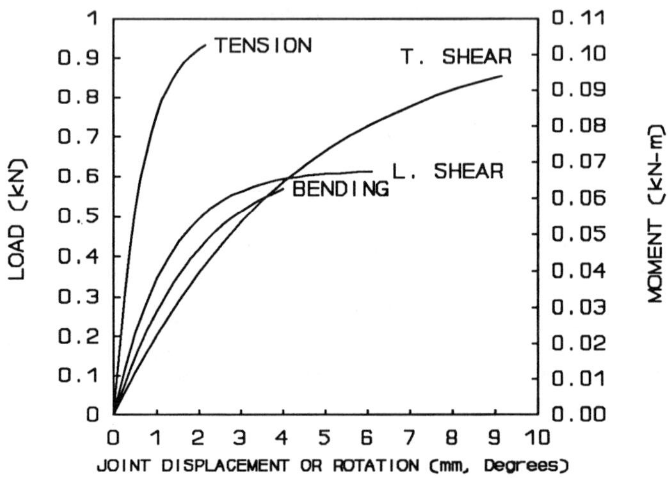


FIG. 4—Curves fitted to single joint test data.

block long specimens were tested (Fig. 5). They were fabricated by stacking the blocks with 197 mm (7.75 in) lengths of spacer, then gunning and tooling the joints. One set used the spacer supplied by the manufacturer and a second set had the notched spacer. The ends of the specimen were supported by snap-fit plastic channels supplied by the block manufacturer. These appeared to behave essentially as simple supports. There were three specimens per set. The test speed was 1.3 mm/min (0.05 in/min). The bending and shear stresses in these tests are very much like those that would be experienced by a long, narrow panel in which the wind load is carried primarily by the short direction, in what is commonly referred to as one-way bending. Typical results for the standard spacer are shown in Fig. 6. All failures were by cohesive tearing of the bead on the tension face of the center joint. One specimen did exhibit significant shear displacements in the joints between the loads and supports, but still failed in tension at the center joint. The notched spacers gave an average strength 27 percent higher that for the standard spacer.

## Two-way Bending Test

One test specimen was constructed and tested to evaluate two-way bending response under a simulated wind load. In conventional plates two-way bending is significant when the ratio of adjacent panel sides is between 1:2 and 2:1. In these panels, significant bending stresses are present in both the horizontal and vertical directions. The test specimen was a square panel, with six blocks on a side. The support frame was built from 38 mm by 184 mm (nominal 2 x 8) lumber. Lengths of snap-fit plastic channel were securely attached by screws to the bottom and sides of the frame. The horizontal courses of blocks were fabricated outside of the frame in a manner similar to that used for the one way specimens. The silicone in these courses was allowed to cure for a week at ambient room conditions, averaging 21° C (70° F) and 47% RH. Then the first course was press fit into the bottom channel. A continuous layer of the vinyl spacer was placed atop the first course and then the prebonded second course was placed on the spacer. This process was continued until all six courses had been placed. Then the top of the frame with its plastic channel was attached. The horizontal beads were gunned and tooled and the entire panel was allowed to cure for two months at ambient room conditions prior to testing. It should be noted that the manufacturer's intended procedure was to dry fit all blocks and then gun and tool all joints. We deviated from this to make it easier to work with the bare support frame, which normally would be tied into adjacent wall elements.

In order to induce a uniform load on the panel, the air bag method was used. A plastic bag, with a small fill tube in one corner, was attached to a piece of heavy plywood reinforced with lumber. The plywood was attached to the rear of the panel support frame so that the air bag was in a narrow space formed by the glass block panel, the frame and the plywood. Fig. 7 shows a schematic of the panel and test setup. A gage bar was made to hold six dial gages and aluminum support brackets were secured to the frame at the center of each horizontal course of blocks. By holding the gage bar on each pair of brackets, an initial zero deflection reading was taken at the center of each block. After these initial readings, the air bag was

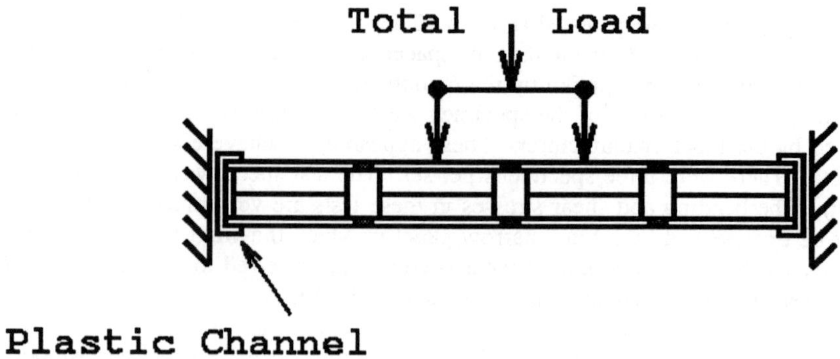


FIG. 5—Set-up for one-way panel test.

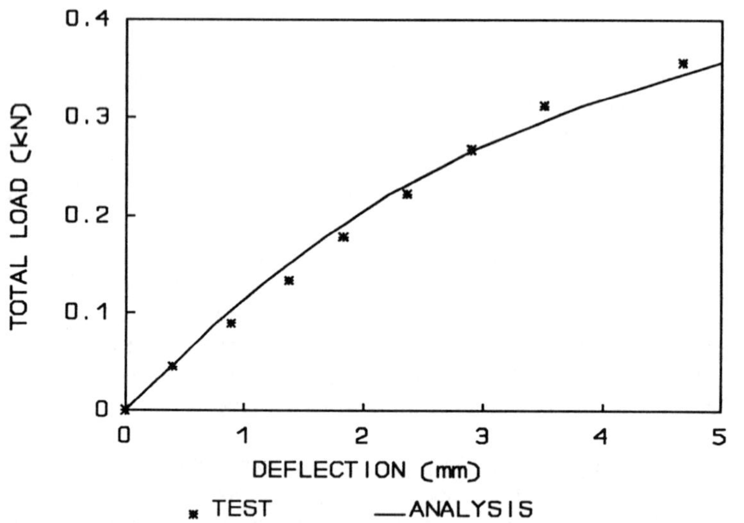


FIG. 6—Average of one-way tests vs. finite element analysis.

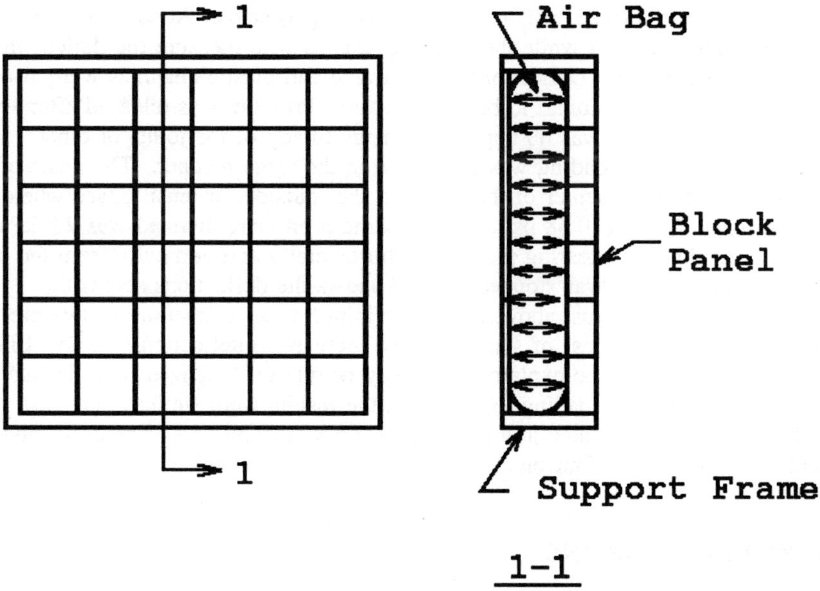


FIG. 7—Schematic of the two-way test panel.

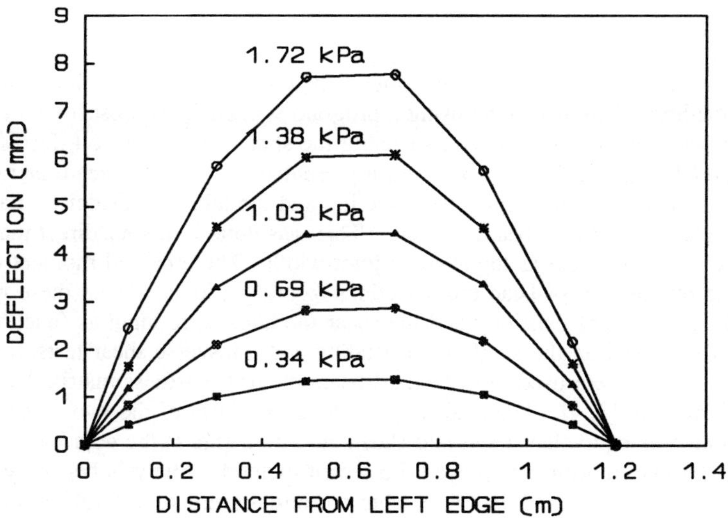


FIG. 8—Load vs. deflection, near the panel mid-height.

pressurized to 1.7 kPa (0.25 psi) in 0.34 kPa (0.05 psi) steps. At each step the pressure was held constant while the gage bar was used to measure the deflections at the centers of the blocks. Deflections of the frame and plastic channels were also checked at each step and found to be insignificant. Pressure was released after reaching 1.7 kPa. There was no apparent damage to any of the joints or other components. A second loading was conducted in the same manner. The onset of tearing was noted at the center of the panel on the outside, or tensile, face when the pressure reached 2.6 kPa (0.38 psi). The maximum pressure attained was 2.9 kPa (0.42 psi). At this load, tearing propagated horizontally and vertically from the center of the panel and the test was stopped. Fig. 8 shows the deflections across the horizontal row of blocks just above the center joint line as a function of pressure for the first loading. Some loss of symmetry and increase in deflection was noted in the second test. This cannot be explained, but may be due to localized joint damage sustained during the first loading. Otherwise, the results were similar. The last deflection readings were taken at a pressure of 2.5 kPa (0.36 psi). The deflection at the centers of the middle four blocks averaged 14 mm (0.55 in).

## ANALYTICAL MODELING

For silicone bonded glass block panels, like any other structural system, analytical models verified against test data provide the most efficient approach to design. Toward this end, a full load range model of one-way panel behavior was developed. A preliminary model for two-way panels under service loading was also investigated.

### One-way Bending Model

The analysis of the one-way case was approached using beam elements in the finite element method. This was done using a program written in Microsoft QuickBasic. A strip of blocks, one block wide and spanning the short panel direction, forms the basis for the model (Fig. 9). The blocks were represented by very stiff beam elements based on the actual block length, cross section and modulus of elasticity. Each sealant joint was also treated as a beam element. This was done in a very direct manner, taking the element length as the average joint width. The area and moment of inertia were those for the pair of bead cross sections forming a joint. Given these geometric values, the modulus of elasticity and the shear modulus were input as functions of displacement, based on the single joint bending and transverse shear tests, respectively. These parameters were used in a standard beam element stiffness matrix that includes shear deformation [3, 4]. The loads applied to the middles of the two inner blocks were modeled as equivalent fixed end shears and moments at the adjacent nodes. The iterative solution algorithm employed the secant method. Analysis began by applying the full value of the desired load and using the initial (zero strain) values for E and G. The program then solved for nodal displacements and sealant element stress resultants in shear and bending. The nodal displacements were used to calculate updated, reduced values of E and G for another solution. This was repeated until the values of the updated stresses for each joint were within 0.1 percent of the stresses from the

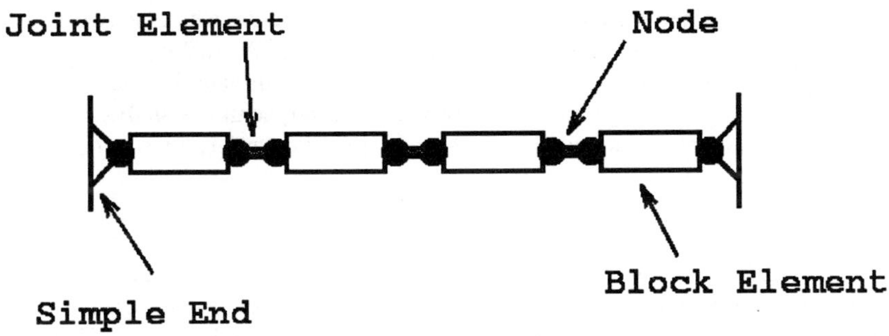


FIG. 9—One-way bending finite element model.

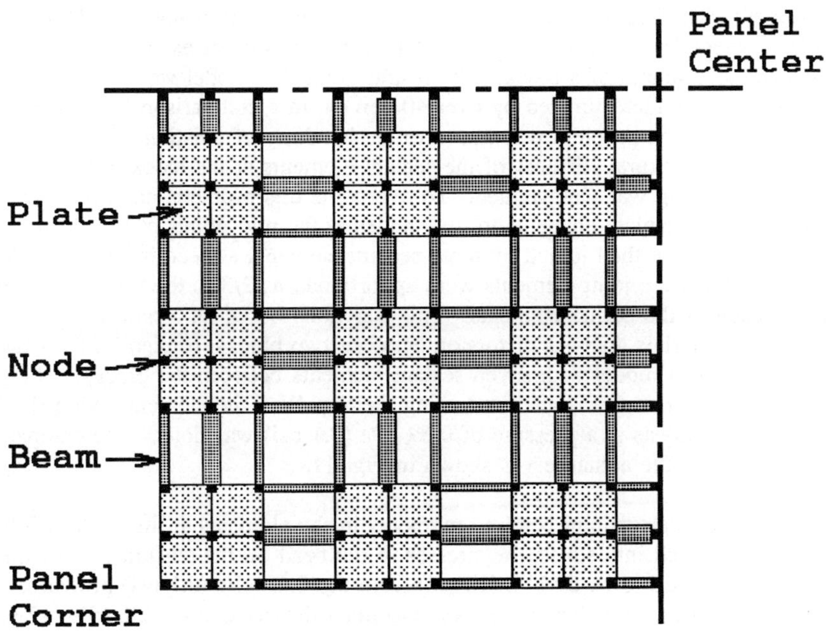


FIG. 10—Two-way panel finite element model.

previous iteration. When all joints met this limit, convergence was assumed and the final displacements and joint stress resultants were output. By making a series of analyses at progressively higher loads, it was possible to simulate the four block, one-way test. In Fig. 6, the comparison is fairly good overall, with the analysis being somewhat stiff at lower loads and slightly soft as ultimate load is approached.

Two-way Bending Model

Panels exhibiting two-way bending require a plate type of model. A silicone bonded glass block panel is an unusual plate-like structure because it consists of the essentially rigid blocks interconnected with the narrow but highly flexible silicone joints. Our approach to this problem was to extend the one-way model to two-way action. In addition to bending in two directions, plate behavior involves a twisting action which is most significant near the corners. The blocks were modeled with standard four node shell elements. The pressure was modeled with transverse nodal loads based on the tributary area for each node. The snap-fit channels were treated as simple edges and quarter symmetry was used. In our first modeling attempt we used one element per block and connected any two adjacent blocks with two beam elements at the corners. The beam elements modeled the sealant in the manner used for the one way joints except for the fact that the tributary bead lengths for each element were only half the side length of a block. The results from this model were not good. Deflections were underestimated by over 40 percent in a comparison to the 6 by 6 block test panel. The problem was traced to the handling of twisting action. By concentrating the flexural stiffness of the sealant elements at the block corners the resistance to twisting was exaggerated. To overcome this problem, each block was subdivided into four plate elements to give nodes at the mid points of the sides. Adjacent blocks were then joined by three sealant elements as seen in Fig. 10. The stiffnesses of the three joint elements were apportioned as 2/3 to the mid side element and 1/6 to each of the corner elements. This was done based on a rudimentary mechanics of materials analysis of torsion between two blocks, backed up by a two block finite element model with seven sealant elements between the blocks. The resulting two way model was analyzed using the Sap IV finite element code [5]. Only a linear elastic analysis at a pressure of 0.69 kPa (0.1 psi) was done. The comparison of deflections with the actual test is shown in Fig. 11.

When the sealant joint bending moments in the elements at the center of the panel were converted into stress, the stresses in the bead on the tension face of the panel were calculated to be 69 kPa (10 psi) from the 0.69 kPa (0.1 psi) pressure. This implies that the stress at failure in the test specimen did not exceed 290 kPa (42 psi) Initially we were troubled by this apparent low strength value, especially since the joints in the center of the panel had the proper bite, no voids, and failed cohesively. A likely explanation is that there is a significant stress concentration where the horizontal and vertical beads intersect. At the very center of the panel, the tensile face corners of all four blocks move away from the center along their respective diagonals. A separate finite element analysis of the localized sealant stresses at the intersection suggests maximum tensile stress in the diagonal direction of over 1.7 times the bending stress calculated in the analysis of the panel. This could readily account for the apparent low stress at failure.

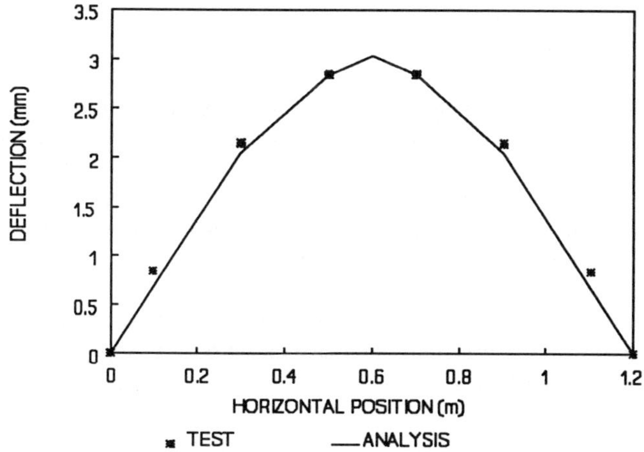


FIG. 11--Two-way test results vs. finite element analysis at 0.69 kPa pressure.

## DESIGN CONSIDERATIONS

The four block wide, one way panel is limited by strength considerations. The load required to produce a deflection of L/180 is approximately 90% of the ultimate strength. With any reasonable factor of safety, the design load would be reached before deflection became a problem. Applying a 2.5 factor of safety to the ultimate bending strength leads to an allowable design pressure of 1.4 kPa (30 psf) for a four block wide panel.

The design load for the six block wide square panel is also limited by strength rather than stiffness. Applying a 2.5 factor of safety to the 2.9 kPa ( 60 psf) ultimate pressure, gives a possible allowable design pressure of 1.2 kPa (24 psf).

It appears then, that for moderate panel dimensions and design wind loads, it would be possible to produce acceptable exterior designs using silicone sealants to bond the blocks. Note that the design values here are purely hypothetical and are based on extremely limited data.

## CONCLUSIONS

The prototype system of blocks, sealant, spacer, and perimeter is feasible from a structural standpoint. The deflections of the one-way specimens and the two way panel would be acceptable for moderate design wind loadings.

The behavior of silicone bonded glass block panels can be modeled successfully using ordinary beam and plate elements, available in commercial structural analysis software.

In two-way panels, stress concentrations in sealant joints at the middle of the panel may be the significant factor limiting the load capacity of the panel.

The large number of joints in a glass block panel suggests the use of a very high strength, high stiffness sealant. The ideal sealant may be one which is actually too stiff for conventional structural glazing applications.

There is definite potential for an improved spacer which more readily aligns and holds blocks during assembly, and sealant cure. An improved spacer should also allow for greater joint bite.

## ACKNOWLEDGEMENTS

This study was funded in part by an unrestricted grant from the Dow Corning Corporation. Blocks and accessories were provided by the Pittsburgh Corning Corporation. The authors gratefully acknowledge this support.

## REFERENCES

[1] Chang, K. F. "Structural Behavior of Adhesive-Bonded Joints in Glass Block Panels," dissertation presented to Michigan Technological University for the degree of Doctor of Philosophy, 1994.

[2] Sandberg, L. B. and Tan, I. P. "A Load-Displacement Model for Structural Glazing Joints in Tension and Shear," Science and Technology of Building Seals, Sealants, Glazing, and Waterproofing, ASTM STP 1168, C. J. Parise, Ed., American Society for Testing and Materials, Philadelphia, 1992, pp. 85-96.

[3] Ghali A. and Neville, A. M. Structural Analysis, Intext Educational Publishers, New York, 1972, pp. 400-402.

[4] Weaver, W. Jr. and Gere, J. M. Matrix Analysis of Framed Structures, D. Van Nostrand Co., 1980, pp. 411-413.

[5] Bathe, K-J., Wilson, E. L. and Peterson, F. E. "SAP IV A Structural Analysis Program for Static and Dynamic Response of Linear Systems," National Technical Information Service, 1973.

# Aesthetic Performance of Sealants

Virginia K. O'Neil and Andreas T. Wolf[1]

# EFFECTS OF WEATHERPROOFING SEALANTS ON BUILDING AESTHETICS - PART II

---

**REFERENCE:** O'Neil, V.K. and Wolf, A.T., **"Effects of Weatherproofing Sealants on Building Aesthetics–Part II"**, Science and Technology of Building Seals, Sealants, Glazing, and Waterproofing: Fifth Volume, ASTM STP 1271, Michael A. Lacasse, Ed., American Society for Testing and Materials, 1996.

**ABSTRACT:** The paper describes several sealant related phenomena which can contribute to building aesthetic problems. These include sealant dirt pick-up, fluid streaking on a substrate (staining), and fluid migration into porous substrates. The paper postulates mechanisms for these phenomena and suggests appropriate test methods to assess them. The paper also presents long-term weathering results obtained on 12 weatherproofing sealants selected from the following generic types: plasticized silicones, unplasticized silicones, unplasticized surface-modified silicones, plasticized polyurethanes, unplasticized polyurethanes and plasticized silicon modified polyethers. In order to address the influence of local environment and exposure conditions, sealants are exposed at a rural site in Georgia and an industrial site in Michigan.

**KEYWORDS:** aesthetics, fluid streaking, dirt pick-up, fluid migration, chalking, sealants

---

The appearance of any building degrades with time, this is the result of complex processes generally summarized under the term "aging". Some of the appearance changes are associated with a non-uniform darkening of the building substrate, a phenomenon generically referred to as "staining". As has been pointed out recently [1], this staining is well accepted, and actually valued, by many of us when considering historic buildings, such as Japanese temples or shrines. For modern buildings, however, staining is considered a nuisance, since it negatively affects their real estate or rental value. According to

---

[1]Technical Service and Development Chemist and Global Technology Manager, respectively; Sealants Science and Technology, Dow Corning Corporation, Midland, MI 48686.

[1], there are two reasons for the widely varying attitude towards building staining: First, the different value we place on aging when associated with historic versus modern buildings; second, the fact that historic buildings are generally owned by the public, while modern buildings are mostly privately held.

## EFFECTS OF SEALANTS ON BUILDING STAINING: DEFINITION OF TERMINOLOGY AND DESCRIPTION OF PHENOMENA

Sealants can contribute to the degradation of a building's appearance in various ways. However, it is important to understand that many forms of substrate defacement will appear over time with or without the presence of a sealant. Another problem is that there is no uniformly accepted terminology for the various forms of substrate appearance changes. For this paper, the terminology for the three classes of sealant related aesthetic phenomena will be used as defined in a previous paper on this topic [2].

### Fluid Migration

The term refers to the migration of fluids (plasticizers, unreacted or unreactive polymers, antioxidants, or other liquid formulation additives) out of a sealant and into a porous or microporous substrate causing a dark band ("wet look") around the joint. The fluid soaked band may extend from a few millimeters to several centimeters from the joint, depending on the amount and the viscosity of the material bled from the sealant. Migration of any fluid, regardless of its nature, will cause a "wet look" to occur on dry substrates, whereas hydrophobic fluids will, in addition, impart a "dry look" to substrates under wet or humid conditions. Porous stones, such as granite or marble, that have been soaked with migrating fluids usually do not attract and retain contaminants ("dirt"), except when excessive fluid migration causes formation of a surface film or when the migrating fluid is hydrophobic and of high(er) molecular weight.

The extent and rate of fluid migration into porous substrates depend primarily on the sealant's formulation and secondly on its cure chemistry. Contrary to common belief, the chemical nature of a sealant's polymer has little or no effect on the degree of fluid migration observed. Consequently, fluid migration has been observed for any sealant type: oleo-resinous putties, butyls, acrylics, polysulfides, polymercaptanes, polyurethanes, silicon modified polyethers, and silicones. As an example, Fig. 1 shows severe fluid migration from a low modulus, one-part polyurethane sealant into a marble substrate. Both the extent and rate of fluid migration depend on the viscosity of the materials bled from the sealant; the lower the viscosity of the migrating fluid(s), the higher the diffusion rate and the extent of the fluid migration are. A "wet look" caused by low viscosity plasticizers, for example, will therefore appear faster and extend farther into the substrate, than if high molecular weight plasticizers were used. There are two disadvantages to plasticizing with high molecular weight fluids. First, once the high molecular weight plasticizer diffuses into the porous substrate (and eventually it will!), it causes a very unsightly stain due to the resulting dirt pick-up on the fluid soaked substrate. Second, the diffusion rate of higher molecular weight fluids is often accelerated by the presence of a lower molecular weight species in the sealant that acts as a "carrier fluid".

FIG. 1 -- Severe fluid migration from a low modulus, one-part polyurethane sealant into a marble substrate (shopping mall near Birmingham, United Kingdom)

FIG. 2 -- An anodized aluminum cladding exhibits fluid migration patterns caused by a benzamide cure, low modulus silicone sealant (Bad Kreuznach, Germany)

Substrate porosity and color also are key factors in the extent and visibility of fluid migration. The finer the porosity of the substrate, the higher the rate and extent of fluid migration. Light colored substrates, such as marble, limestone and granite, are very prone to discoloration by fluids. However, fluid migration can occur on substrates other than natural stones. Fig. 2 shows, as an example, fluid migrated from a low modulus, one-part benzamide cure silicone sealant onto and into a bronze colored, anodized aluminum causing a "picture framing" effect. The fluid migration patterns on porous stone and anodized aluminum are similar, despite the difference in substrates, because fluid migration is a diffusion controlled phenomenon.

Dirt Pick-Up

The term dirt pick-up describes the accumulation of airborne particulates on the sealant surface, though it has been applied to fluid contaminated substrate surfaces that are soiled. Unless otherwise noted, the term "dirt pick-up" as used here will refer to the sealant surface. Dirt pick-up may occur during or after sealant cure. Sealants with slow surface cure (long tack-free time), such as one-part polysulfides or polyurethanes, are especially prone to pick up dirt on dusty construction sites. Dust that is embedded in the sealant surface during cure is inextricably bound and cannot be removed without abrading away the sealant surface.

There appear to be three mechanisms for dirt pick-up on cured sealant surfaces: residual tack, fluid migration, and physical adsorption. First, if a cured sealant has high residual tack, dust particles will adhere to the sealant surface by the same mechanisms as pressure-sensitive adhesives stick to substrates. Continued dust accumulation and weathering eventually cause the sealant surface to become tack-free. As is the case with dirt pick-up during cure, dirt that attaches to a sealant due to residual tack cannot be removed by cleaning the surface. Residual surface tack may be caused by unreactive or unreacted polymer, which in turn can result from insufficient cure as observed for inhomogeneously mixed two-part sealants or some out-of-shelf-life one-part sealants. Surface tack may also be caused by polymer reversion, as has been observed for some polyurethane sealants [3]. Second, sealants exhibiting fluid migration into porous substrates generally also have a surface film which retains loose dirt. Fig. 3 shows, as an example, severe dirt pick-up on the surface of a one-part, medium/low modulus, alkoxy cure silicone sealant. Note that the dirt film can be easily scraped off. The authors suggest that the mechanism by which dirt attraction and retention appear to occur in this case is as follows: dirt particulates are deposited on the fluid film and over time become completely overcoated by the fluid. Since most plasticizers and other migrating fluids are hydrophobic in nature, the immersed dirt cannot be washed off by rain; therefore, a thick, oily film forms on the sealant surface with time. Generally, this type of surface contamination can be removed by surface cleaning, although a mild detergent or solvent may be required. The third mechanism of dirt pick-up, physical adsorption, is the least understood. A commonly held belief is that this mechanism is somehow related to the static electric surface charge [4]. Recent findings [5] on sealants containing surfactants and/or electroconductive additives have shown that the static electric surface charge plays only a minor role. Physical adsorption appears to be directly related to the viscoelastic properties of the sealant's polymeric network [6,7]. A good rule of thumb is that the lower a sealant's surface hardness, the higher its potential for dirt pick-up is.

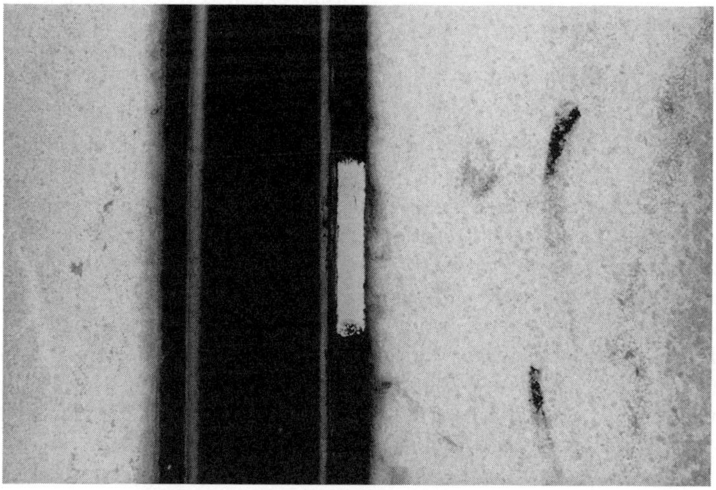

FIG. 3 -- Severe dirt pick-up on the surface of a one-part, medium/low modulus, alkoxy cure silicone sealant that can be easily scraped off (high-rise building, Chicago)

FIG. 4 -- Mold growth and subsequent dirt pick-up on a polyurethane sealant which cannot be removed by physical cleaning (high-rise building, Hongkong)

Although the mechanisms for dirt pick-up are mostly independent of polymer type, dirt retention is related to the weatherability of the sealant surface. A less durable sealant surface will erode away over time; this erosion will also remove any dirt attached to the surface. Finally, it should be noted that microbial growth on a sealant surface can contribute significantly to dirt pick-up [8]. Fig. 4 shows, as an example, a polyurethane sealant with patches of mildew growth and subsequent dirt pick-up.

Dirt or Residue Run-Down

Although not sealant related, dirt (or residue) run-down is briefly discussed here to clarify its cause. Dirt run-down refers to the vertical streaking observed on buildings that is caused by dirt collecting on roofs, window ledges, or other horizontal areas and being channeled down the facade by water run-off. Dirt run-down occurs with or without the presence of sealants. Fig. 5 demonstrates an example of dirt run-down occurring on a facade in the absence of a sealant. Joint design, as well as differences in the wetability of the building facade, may affect the water run-off patterns. Since horizontal building joints are areas in which dirt can accumulate, vertical streaks are frequently observed below these joints and are often mistakenly attributed to the presence of the sealant. Residue run-down can also be caused by materials leaching from the substrate or other non-sealant building materials (e.g. efflorescence from masonry).

FIG. 5 -- Intense dirt run-down pattern visible on non-porous panels in the absence of a sealant (O'Hare airport, Chicago)

Fluid Streaking

The term fluid streaking was coined by these authors to characterize the accumulation of airborne particulates in areas where fluid has migrated from the sealant onto non-porous substrates and forms discreet channels on the building facade. On first glance, fluid streaking and regular dirt run-down look quite similar and are often mistaken for each other. However, fluid streaking, when removed by physical cleaning, will generally recur much faster than regular dirt run-down. This faster recurrence is due to remnant fluid on the substrate that is usually not removed by ordinary cleaning techniques. Furthermore, fluid streaking often is visible above and below horizontal joints, while dirt run-down typically is observed only below horizontal joints.

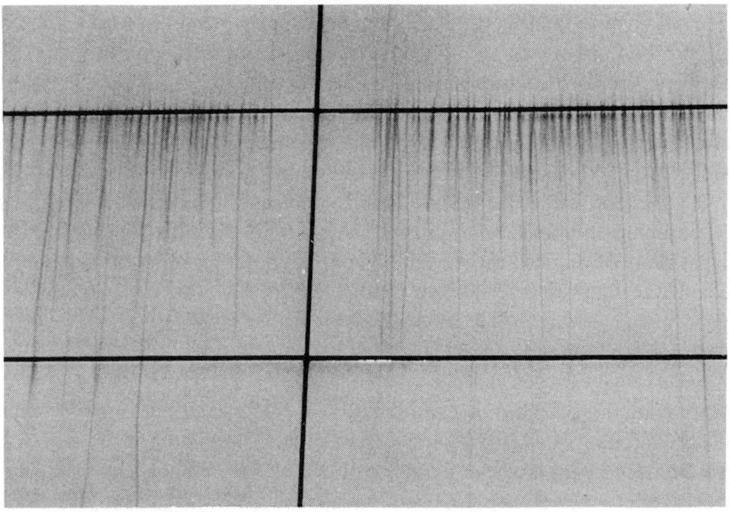

FIG. 6 -- Severe fluid streaking caused by a neutral cure silicone sealant applied between enamel coated cement panels (Atlanta)

While migration of free fluid to a sealant surface is a necessary requirement for fluid streaking, it is incorrect to assume that a sealant not showing fluid migration into porous substrates would not have the potential for fluid streaking. The difference between the two phenomena is that, in order for them to become detectable, fluid migration into porous substrates requires the migration of much more free fluid than necessary for fluid streaking. Quite small amounts of hydrophobic fluids on an essentially non-porous surface are effective in retaining dust particles in water run-down channels. The inverse deduction, however, holds true: sealants that show fluid migration into porous substrates have the potential to show fluid streaking. Whether or not they actually will cause fluid streaking depends on the type of substrate and the type of free fluids involved.

As with fluid migration, fluid streaking can occur for any type of sealant regardless of the nature of its polymeric backbone. Plasticizer bleed, for instance, gives rise to easily detectable fluid streaking on substrates such as reflective coated glass, ceramics, polyvinylidene difluoride coated metals, and enamel coated panels. Fig. 6 shows, as an example, severe fluid streaking caused by a neutral cure silicone sealant applied on enamel coated cement panels.

The duration of fluid streaking depends on the sealant formulation and the water solubility and weatherability of the fluid. The higher the level of free fluids in the sealant formulation, the faster the phenomenon will occur and the longer it will last. Most free fluids bled from sealants are hydrophobic in nature and have poor water solubility. However, some organic fluids are of poor weatherability and are degraded to lower molecular weight species, which can be solubilized or at least emulsified in water, and are subsequently washed off with rain. Nevertheless, rainwater can act as a physical force to spread the fluids, making the fluid streaking highly visible before degradation can occur. Some organic plasticizers, such as dioctylphthalate (DOP), readily degrade when exposed to the combined action of ultraviolet light and humidity. Other organic plasticizers, such as chlorinated paraffines, are environmentally very stable. Silicone plasticizers are very weather resistant as well. Organic sealants containing weatherable fluids tend to show cyclic fluid streaking [2]. Regardless of fluid type, diffusion is increased in summer, when the sealant is compressed and when higher temperatures increase fluid mobility. During winter, when little fluid migration is occurring, the environmental degradation of some organic fluids on the substrate is faster than the replenishing of fluid.

Chalking

The term chalking refers to the whitening of some organic sealant surfaces with weathering. The whitening is caused by the degradation and erosion of organic polymer, which leads to increased roughness or porosity of the sealant surface. Some sealant formulations weather so badly that all organic material leaches from the surface, leaving behind nothing but the inorganic fillers, typically calcium carbonates, hence the term. Chalking turns sealants which, for instance, are initially black or brown into non-uniform light grey or beige sealants.

Although chalking does not occur in acrylic or silicone sealants, it is, however, a common occurrence with polyurethane sealants [9]. Certain polyurethane sealant products are prone to "white run-down", a phenomenon generally believed to be caused by rain water washing off the chalked surface of these products [10]. To a lesser extent, chalking has been observed in polysulfide sealants, especially those formulated in brown or bronze colors. This is not only because chalking is most easily detectable on these colors, but apparently also because some pigments act as photosensitizers [10]. The inclusion of antioxidants and ultraviolet light absorbers, commonly found in most organic sealants, can delay the deterioration of the surface. Even with careful formulation and the use of organic based polymers with a reputation for improved durability, such as silicon modified polyether sealants, chalking can occur (Fig. 7).

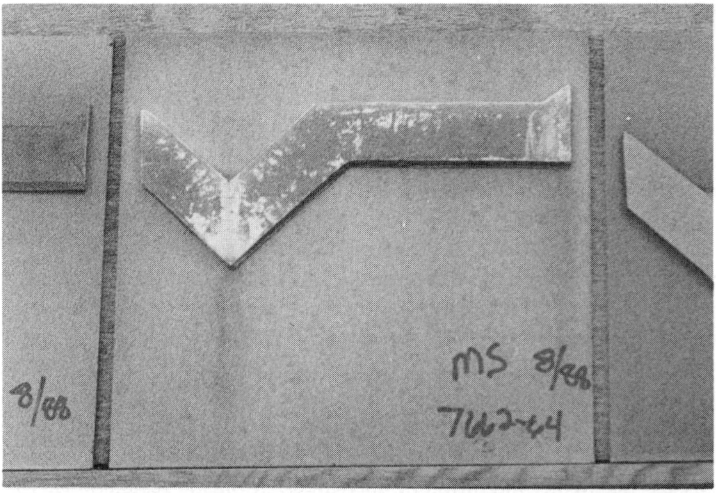


FIG. 7 -- Severe chalking and dirt accumulation are visible on this low modulus, neutral cure silicon modified polyether sealant (photo taken in 1994 after 6 years weathering).

The increased porosity of the sealant surface caused by the deterioration allows dirt particulates and mold spores to settle more easily. Any biocide at the surface is also eroded away with the polymer matrix. In addition, some degraded polymers can serve as nutrients for microbes. These effects synergetically increase the chance for mold fungus growth on chalking organic sealants. In cases where mold is not substantially present, chalking can act as a cleansing mechanism to remove sealant dirt pick-up. However, a sealant that is lighter in color, versus darker due to dirt pick-up, still presents an aesthetic issue.

## TEST METHODS ASSESSING THE EFFECTS OF SEALANTS ON BUILDING AESTHETICS

As elastomeric sealants were developed and commercialized during the last 30-40 years, the primary concern of sealant developers and specifiers was the actual weatherproofing performance of these materials. Numerous test methods for adhesion, movement capability, and water resistance were developed, focused solely on the functional performance of sealants. However, as the utilitarian side of sealants improved over time, the effects sealants have on the aesthetic appearance of buildings were increasingly being considered by building owners, architects and specifiers. While this is a recent trend, explaining why only few standardized test method exist, it has become so strong that aesthetic considerations now often override the practical considerations in the sealant selection process. It is therefore highly desirable to develop standardized test methods that assess the overall performance of sealants. The following test methods have been successfully used by these authors to assess the aesthetic performance of sealants.

## Fluid Migration

The recently published ASTM Test Method for Staining of Porous Substances by Joint Sealants (C1248-93) [11] is a useful test method for sealant manufacturers to check the fluid migration potential of a newly developed sealant. The test consists of subjecting cured standard joint specimens, under compression according to the manufacturer's rated movement capability, separately to heat, UV light with alternating water condensation, and ambient exposure conditions for a total of one month. After this conditioning, the specimens are inspected for any signs of fluid migration on the surface and through the cross-section of the substrate. While this test method is useful for sealant manufacturers, who know the formulation of the sealant they study, it can be misleading, if used by someone not having this information. High molecular weight plasticizers and unreacted polymers, which take considerable time to migrate from the cured sealant, are not detected by this test method. A potential upgrade may be to expose the cured specimens in a 50% compressed state at 70°C for 56 days or 74 days; although this timing would seriously limit the use of this test as a sealant selection technique for most projects. Tests that were developed to assess plasticizer migration of polysulfide or plastic sealants [12-14] cannot be used to assess the long term fluid migration potential of sealants in general.

## Dirt Pick-Up

While significant efforts went into the development of an accelerated test method to assess the dirt pick-up of sealants, the authors realize that any accelerated method is flawed with artifacts, since any "artificial dirt" does not have the same composition as the airborne particulates present in our environment. Furthermore, efforts to reproduce the deposition and retention of dirt on sealant surfaces with some time acceleration have failed. Thus, the best method for assessing the dirt pick-up potential of a sealant still is real-time outdoor exposure at a 90° angle of inclination. Even in outdoor exposure tests the composition of the airborne dirt particulates and the factors influencing dirt deposition and retention will vary between exposure sites and exposure conditions (geographical location, nature of environment, compass direction of exposure, inclination of specimens). Outdoor exposure tests, therefore, can only be used for comparative purposes by exposing, for example, a series of different sealants under identical conditions at different sites.

## Fluid Streaking and Dirt Run-Down

Initial efforts by these and other researchers [15,16] utilized a V-shaped or "square root" shaped sealant bead applied to reflective glass or enamel coated cement panels in an effort to accelerate fluid streaking in outdoor weathering test. This method was, however, flawed with poor reproducibility, since natural dirt run-down often masked the fluid streaking contribution. A better method is to expose test specimens assembled from glass or other substrates (total dimension: 0.25 x 0.25 $m^2$) with a horizontal sealant joint (25 mm wide x 6 mm deep) in a weathering rack at a 45° angle against the horizontal [2]. The acceleration factor, if any, is unknown, but this method provides good reproducibility of the fluid streaking phenomena. In order to differentiate between dirt run-down and the fluid streaking contribution, it is worthwhile to include a sealant-free control specimen in the test, realizing though that the water channeling over the control specimen is different from that of the sealant containing specimens. The same dependency on environmental and exposure conditions exists for fluid streaking as for dirt pick-up; the proposed test method therefore should only be used for comparative purposes.

Chalking

Chalking of organic sealants can be accelerated by exposing them to ultraviolet light, heat and humidity in a weatherometer for at least 2000 hours. The recommended test duration is based on earlier findings that 2000 hours of exposure in a Q-UV weatherometer corresponds to about 1-2 years of outdoor exposure in the USA [3, 17].

## RESULTS OF OUTDOOR WEATHERING OF COMPOSITE PANELS TO SIMULTANEOUSLY ASSESS FLUID MIGRATION, DIRT PICK-UP, FLUID STREAKING AND CHALKING

The aesthetic performance of twelve weatherproofing sealants were compared by exposing them in a Northern, industrial environment in Midland, Michigan (USA) and in a Southern, rural environment near Atlanta, Georgia (USA). The sealants were selected from the generic types of silicone (plasticized, unplasticized, and unplasticized surface-modified), polyurethane (plasticized and unplasticized) and silicon modified polyether (plasticized). The sealants were installed into composite panel specimens consisting of white Italian marble, fiber reinforced cement panel (uncoated) and reflective glass. The composite was arranged to have the marble in the upper right corner, the fiber reinforced cement panel in the upper left corner, and the reflective glass across the bottom, forming an inverted T-joint between the panels with an average joint width of 25 mm. The entire composite measured 250 x 250 $mm^2$ and was positioned on a weathering rack at a 45° angle facing South.

The composite specimen configuration was selected on its ability to show fluid migration (marble and uncoated fiber reinforced cement panel), fluid streaking (reflective glass) and dirt pick-up on the sealant. However, as was found by comparing the dirt pick-up results with those of a separate study on vertically exposed samples (see below), positioning the samples at a 45° angle to the horizontal accelerates both the dirt deposition phenomena and the sealant degradation rate and, thus, favors sealants with high surface erosion, while penalizing sealants with good durability. Two other limitations of the chosen composite specimen configuration were found during the study: The first limitation apparently is the alkalinity of the water run-down from the uncoated, reinforced cement panel which has an effect on the fluid streaking observed on the reflective glass panel. For most sealants, regardless of their chemical nature, the streaking in the area below the cement panel was more pronounced than in the area below the marble substrate. Since the mechanism by which the alkalinity contributes to the increased fluid streaking on reflective glass could not be resolved, nor its relevance for practice be demonstrated, the authors chose to neglect this finding as an artifact and, for the comparative purpose of this study, to only consider the fluid streaking observed below the marble panels. The second limitation is less serious: While the uncoated cement panel shows fluid migration very effectively as a dark "wet" band, it provides less differentiation between moderately and strongly bleeding sealants than the marble substrate does. The authors, therefore, only used the results obtained on marble to quantify fluid migration of the various sealants. Because of the above mentioned limitations, they recommend not to consider uncoated cement panels for future composite specimen studies.

Weathering in a Northern, Industrial Environment

The composite specimens were prepared on 21st and 22nd of July, 1992 and cured four days in a laboratory under ambient conditions (23°C, 55% r.h.) before installing them on an outdoor weathering rack within the fences of a chemical plant in Midland, Michigan, less than one kilometer away from a second chemical plant. Aesthetic results obtained after 3, 10, 18 and 25 months of weathering are displayed in Table 1 and in Figs. 8-20. The three and ten month results are reproduced from [2]. The photographs were taken after 27 months of exposure. The two polyurethane sealants (20-11 and 20-12) show different degrees of fluid streaking after 25 and 27 month exposure; their ratings in Table 1 therefore do not correlate with the fluid streaking shown in the corresponding photographs.

As can be seen from Table 1 and Figs. 8-20, the dirt pick-up of all silicone sealants, whether plasticized or unplasticized, with the exception of the surface modified silicone sealants, continually increases with exposure time. Some silicone sealants pick up dirt faster than other, however, the dirt pick-up on all non surface modified silicone sealants is visible from more than 1.5 meters distance after an exposure time of 25 months (dirt pick-up ratings of 4 or 5 according to [18]). The dirt pick-up on the one-part surface modified silicone sealant (20-7) remains constant, within the accuracy of the measurement, after 10 month exposure time (dirt pick-up rating of 3). The dirt pick-up on the two-part surface modified silicone sealant (20-8) initially remains low (dirt pick-up rating of 2), however, between 10 and 18 months exposure time, the surface of the sealant is overgrown with mold fungus, subsequently causing an increased dirt pick-up rating of 5. The two silicon modified polyether sealants develop very little dirt pick-up over the 25 months exposure at 45° angle (dirt pick-up ratings of 1-2 and 2). The two-part polyurethane sealant (20-11) initially displays significant dirt pick-up (rating of 3-4 after 3 months exposure), however, due to erosion of the surface by weathering, the dirt pick-up is reduced with increased exposure time (rating of 2 after 10 month exposure). However, between 10 and 18 months exposure time, mold fungus develops on the sealant surface below the cementatious panel and subsequently causes increased dirt pick-up. The one-part polyurethane sealant (20-12) also exhibits very low dirt pick-up, however, at the expense of intensive chalking. Between 18 and 25 months of outdoor exposure, the eroded surface is also partially overgrown with mold fungus, which subsequently causes increased dirt pick-up.

TABLE 1 -- Aesthetic testing results for 3, 10, 18, and 25 month exposure in Midland, Michigan (45° angle)

| Sealant | Type | 3 Month Exposure | | | 10 Month Exposure | | | 18 Month Exposure | | | 25 Month Exposure | | |
|---|---|---|---|---|---|---|---|---|---|---|---|---|---|
| | | D/P[1] | F/S[2] | F/M[3] | D/P | F/S | F/M | D/P | F/S | F/M | D/P | F/S | F/M |
| 20-1 | Si/P/1 | 3 | 3 | 2-3 | 4 | 4 | 5-8 | 4 | 4 | 8-10 | 4 | 4-5 | 8-10 |
| 20-3 | Si/P/1 | 3 | 3 | 2 | 3 | 4 | 5-8 | 4 | 4-5 | 5-10 | 4 | 4-5 | 5-10 |
| 20-4 | Si/P/1 | 3 | 3 | 2-3 | 5 | 4 | 4 | 5 | 5 | 3-5 | 5 | 5 | 5 |
| 20-5 | Si/P/1 | 2-3 | 3 | 9-12 | 4 | 5 | 20-40 | 4-5 | 5 | 10-20<br>40+[4] | 4 | 5 | 20<br>40+[4] |
| 20-6 | Si/P/1 | 2-3 | 3 | 8-10 | 3-4 | 4 | 10-20 | 5 | 5 | 10-15<br>20+[4] | 4 | 4-5 | 10-15<br>20+[4] |
| 20-2 | Si/U/1 | 2 | 3 | 0 | 3 | 3-4 | 0 | 3-4 | 4-5 | 3 | 4-5 | 4-5 | 3 |
| 20-7 | Si/U/SM/1 | 2-3 | 2 | 0 | 3 | 2 | 0 | 3-4 | 2 | 0 | 3 | 2 | 0 |
| 20-8 | Si/U/SM/2 | 1-2 | 2 | 0 | 2 | 2 | 0 | 5[5] | 2 | 0 | 5[5] | 2 | 0 |
| 20-9 | MPe/P/1 | 1-2 | 3 | 2-4 | 2 | 2 | 5-10 | 1-2 | 3 | 5 | 2 | 2 | 5-10 |
| 20-10 | MPe/P/1 | 1 | 4 | 4-5 | 1 | 2 | 0-10 | 1 | 2 | 0-10 | 1-2 | 1-2 | 0 |
| 20-11 | Pu/P/2 | 3-4 | 3-4 | 0 | 2 | 2 | 0 | 2-5[6] | 2 | 0 | 1-5[6] | 1 | 0 |
| 20-12 | Pu/U/1 | 1 | 2 | 0 | 1[7] | 1-2 | 0 | 1-2[7] | 1 | 0 | 2[7]-3[5] | 1 | 0 |
| Control | no sealant | ----- | 1 | 0 | ----- | 1 | 0 | ----- | 1 | 0 | ----- | 1 | 0 |

**Footnotes:**

[1] D/P: Dirt pick-up
[2] F/S: Fluid Streaking
[3] F/M: Fluid migration
[4] Very light stain
[5] Mold growth
[6] Mold growth below cement panel
[7] Chalking

**Sealant Types:**

Si: Silicone
Pu: Polyurethane
MPe: Silicon modified polyether
P: Plasticized
U: Unplasticized
SM: Surface modified
1,2: One- or two-part

**Dirt Pick-Up and Fluid Streaking Ratings [18]:**

1 - Sealant is as clean as before exposure and panel is as clean as control without sealant
2 - Sealant is slightly dirty and panel is slightly streaked
3 - Moderate dirt pick-up on sealant and panel is streaked
4 - Dirt pick-up on sealant is visible from >1.5 meters distance and panel is covered with hazy film
5 - Heavy dirt pick-up on sealant and on panel

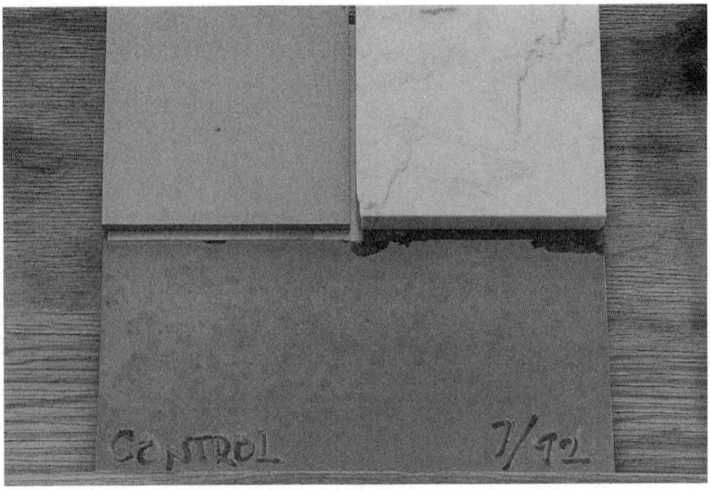


FIG. 8 -- Control composite panel, comprised of white Italian marble, uncoated cement, and reflective glass, in a Northern, industrial site at a 45° angle facing South

FIG. 9 -- One-part plasticized silicone sealant (20-1) (exposure conditions as in Fig. 8)

FIG. 10 -- One-part plasticized silicone sealant (20-3) (exposure conditions as in Fig. 8)

FIG. 11 -- One-part plasticized silicone sealant (20-4) (exposure conditions as in Fig. 8)

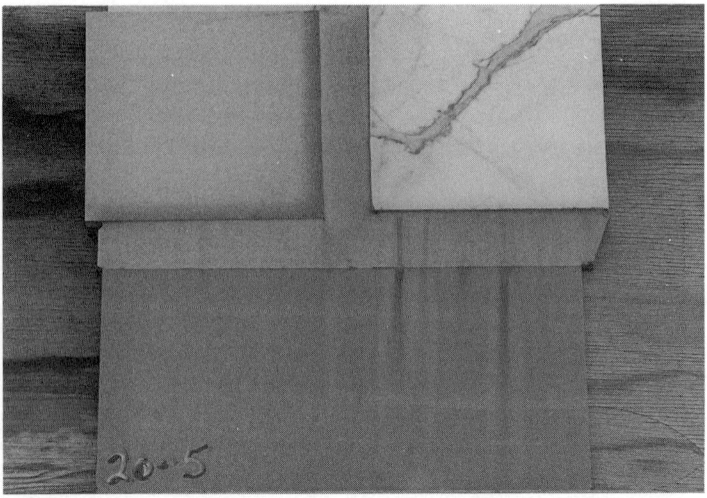


FIG. 12 -- One-part plasticized silicone sealant (20-5) (exposure conditions as in Fig. 8)

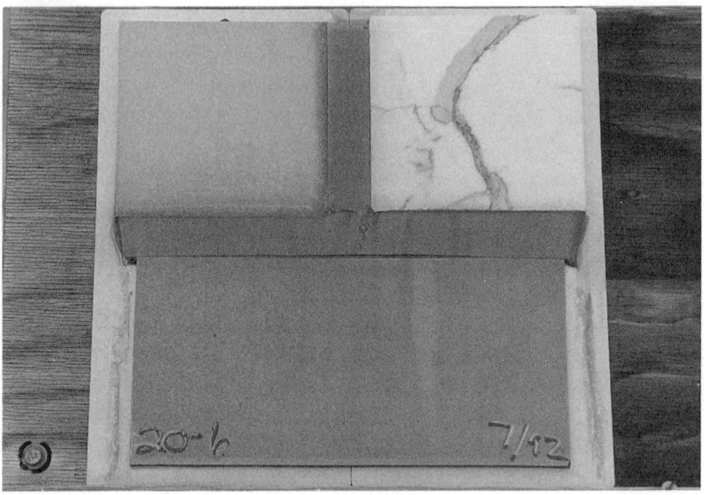


FIG. 13 -- One-part plasticized silicone sealant (20-6) (exposure conditions as in Fig. 8)

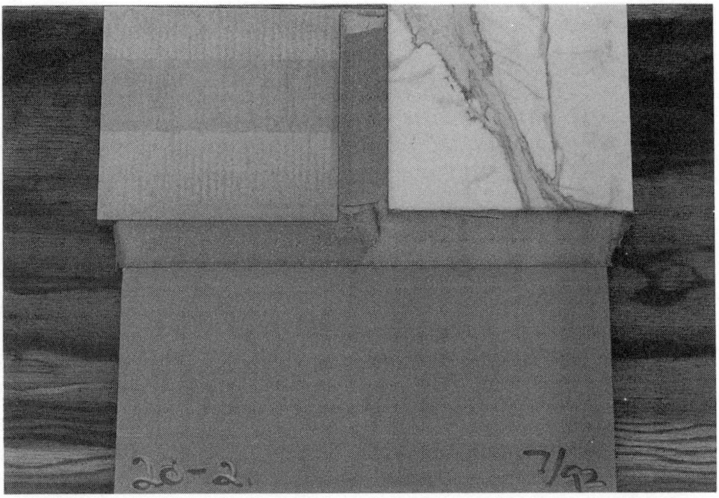


FIG. 14 -- One-part unplasticized silicone sealant (20-2) (exposure conditions as in Fig. 8)

FIG. 15 -- One-part unplasticized, surface modified silicone sealant (20-7) (exposure conditions as in Fig. 8)

FIG. 16 -- Two-part unplasticized, surface modified silicone sealant (20-8) (exposure conditions as in Fig. 8)

FIG. 17 -- One-part plasticized silicon modified polyether sealant (20-9) (exposure conditions as in Fig. 8)

FIG. 18 -- One-part plasticized silicon modified polyether sealant (20-10) (exposure conditions as in Fig. 8)

FIG. 19 -- Two-part plasticized polyurethane sealant (20-11) (exposure conditions as in Fig. 8)

FIG. 20 -- One-part plasticized polyurethane sealant (20-12) (exposure conditions as in Fig. 8)

The fluid streaking of all non-surface modified silicone sealants increases at about the same rate, regardless whether the sealants contain silicone plasticizers (20-1, 20-3, 20-4, 20-5, and 20-6) or not (20-2). After 25 month exposure time, the fluid streaking of these sealants is rated 4 or 5. The fluid streaking of the two surface modified silicone sealants (20-7 and 20-8) remains constant and very low (ratings of 2 after 25 month exposure). The two silicon modified polyether sealants initially show significant fluid streaking on the reflective glass (ratings of 3 and 4 after 3 month exposure), however, after 10 month exposure, the fluid streaking is only slightly visible. For one of the silicon modified polyether sealants (20-9), the fluid streaking increases again after 18 month exposure, while both silicon modified polyether sealants achieve fluid streaking ratings of 2 after 25 month exposure. The plasticized polyurethane sealant (20-11) also displays initially significant fluid streaking (rating of 3-4 after 3 month exposure), however, the fluid streaking diminishes with exposure time. Both polyurethane sealants (20-11 and 20-12) show no fluid streaking after 25 month exposure, but fluid streaking recurs after 27 month exposure (cyclical staining).

Due to variability in the porosity of natural marble, the extent of fluid migration into this substrate varies over the length of the specimen joint. Table 1, therefore, displays ranges between minimum and maximum fluid migration observed on one specimen. With increased exposure time, the fluid migration of some sealants is highly visible within a certain range, and less apparent, but still visible, beyond this range. This phenomenon is described by two ranges in Table 1, such as 10-15 and 20+, meaning that the stain is highly visible up to 10-15 mm in the marble and less apparent up to 20 mm and beyond.

All plasticized silicone sealants show some degree of fluid migration into the marble. Two silicone sealants (20-5 and 20-6) show very strong fluid migration after 3 month exposure; the extent of fluid migration for these two sealants continues to increase up to 18 months of exposure and then remains constant. Both sealants bleed considerable amounts of fluids and the stain is highly visible. While not much differentiation between the fluid migration can be observed for the other three plasticized silicone sealants after 3 month exposure, it becomes increasingly apparent with exposure time that two of them (20-1 and 20-3) have a higher fluid migration potential than the other (20-4). The unplasticized silicone sealant (20-2) does not show any fluid migration up to 10 month exposure time. However, after 18 month exposure, fluid migration up to 3 mm into the marble has developed and is highly visible. With increased exposure time, the extent of fluid migration of this sealant remains constant. The authors suspect that the delayed fluid migration of this sealant is caused by unreacted polymer. It is likely that this type of fluid migration would not have been detected in the accelerated ASTM C1248 test. Both surface modified silicone sealants (20-7 and 20-8) do not show any fluid migration into marble up to 25 month exposure time. Both silicon modified polyether sealants (20-9 and 20-10) initially show significant fluid migration (2-4 mm and 4-5 mm after 3 month exposure time), however, for one of them (20-10), the fluid migration continues to extend and dissipate into the marble, becoming less visible over time. After 25 months exposure, this sealant (20-10) does not show any visible fluid migration anymore. The other silicon modified polyether sealant (20-9) shows cyclical fluid migration, which is strongest after 10 and 25 month exposure. The two polyurethane sealants (20-11 and 20-12) do not show any fluid migration during the exposure time.

Weathering in a Southern, Rural Environment

The composite specimens were prepared on the same dates as the samples exposed in Michigan. The specimens were cured four days in a laboratory under ambient conditions (23°C, 55% r.h.) before express shipping them to Atlanta, Georgia, and installing them on an outdoor weathering rack in a rural area.

The fluid migration results were similar to those observed in Michigan, but the fluid streaking and dirt pick-up results differed significantly. Table 2 shows the aesthetic performance of the sealants after 21 months outdoor exposure in Georgia. As can be seen by comparing Tables 1 and 2, all plasticized silicones showed less fluid streaking on reflective glass after 21 month exposure in Georgia than those weathered 25 months at the Northern industrial site. The most notable differences can be found for two of the plasticized silicones (20-1 and 20-5), which showed only slight to moderate fluid streaking after exposure in Georgia, while these sealants showed severe fluid streaking after exposure in the Northern, industrial environment. The unplasticized silicone sealant (20-2) showed less fluid streaking in Georgia as well, however, since the glass pane broke during transport, the result should be considered with caution. The fluid streaking observed for the surface-modified silicones (20-7 and 20-8) and the organic based sealants (20-9 to 20-12) were consistent with those observed in Michigan.

TABLE 2 -- Aesthetic testing results for 21 month exposure in Georgia (45° angle)

| Sealant | Type[1)] | Dirt Pick-Up[1)] | Fluid Steaking[1)] | Fluid Migration[2)] |
|---|---|---|---|---|
| 20-1 | Si/P/1 | 4[3)] | 2 | 4-6 |
| 20-3 | Si/P/1 | 4[3)] | 3 | 4-6 |
| 20-4 | Si/P/1 | 4[3)] | 4 | 4-5[4)] |
| 20-5 | Si/P/1 | 3[3)] | 2-3 | 35-80 |
| 20-6 | Si/P/1 | 4 | 4 | 20-25[5)] |
| 20-2 | Si/U/1 | 3 | 3-4[6)] | 1-2 |
| 20-7 | Si/U/SM/1 | 3-4[3)] | 2 | 0 |
| 20-8 | Si/U/SM/2 | 4[3)] | 2 | 0 |
| 20-9 | MPe/P/1 | 2 | 1-2 | 8 |
| 20-10 | MPe/P/1 | 1 | 1 | 8-12[5)] |
| 20-11 | Pu/P/2 | 2-3[3)] | 1 | 0 |
| 20-12 | Pu/U/1 | 2[3),7)] | 1 | 0 |
| Control | no sealant | ------ | 1 | 0 |

1) Sealant types, dirt pick-up and fluid streaking ratings same as in Table 1
2) Fluid migration into marble (in mm)
3) Mold growth
4) Dirt collecting on fluid soaked marble
5) Light stain
6) Reflective glass on composite panel was broken during transport, which may affect rating
7) Chalking

Dirt pick-up ratings for the majority of the samples were more severe under the Georgia exposure conditions due to the presence of mold fungus growth. For those sealants that did not develop mold growth under either exposure condition (20-6, 20-2, 20-9 and 20-10), dirt pick-up ratings at both exposure sites were comparable.

As mentioned before, fluid migration into marble for most sealants was similar under the two exposure conditions. Notable differences are the highly plasticized silicone sealant (20-5), which showed significantly higher fluid ingress in a Southern climate, and the two silicon modified polyether sealants (20-9 and 20-10), for which the stain caused by the fluid migration into the marble substrate prevailed for the complete duration of exposure in Georgia.

## DIRT PICK-UP RESULTS ON VERTICALLY EXPOSED SEALANT SAMPLES IN NORTHERN, INDUSTRIAL CLIMATE

Since building joints are rarely oriented at a 45° angle to horizontal, sealant samples were also exposed vertically (at 90° angle) to realistically monitor levels of dirt pick-up. Test specimens were prepared on 9th and 16th of July, 1992 by applying wet sealant on anodized aluminum sheets (150 x 76 $mm^2$) and tooling the sealant paste to a constant thickness of 2.5 mm. Extra care was taken to achieve a smooth sealant surface. After allowing the specimens to cure for 7 days (at 23°C, 55% r.h.), the upper half of each specimen was masked with polyethylene film, and the specimens were installed in a vertical position at the Northern, industrial exposure site, with the lower half of the specimens freely exposed to the environment. For the fifteenth and sixteenth month of exposure, the samples were exposed to the outdoor weather in a horizontal position while the exposure racks were moved from one location to another within the industrial exposure site.

After weathering for certain intervals in the Northern industrial location, the color change or difference in lightness between the masked area and exposed area was determined using a MacBeth Spectrophotometer in accordance with the ASTM Test Method for Dirt Collection Index (D3719) [19]. The dirt collection index is the ratio (given in percent) of the lightness of the exposed sealant area to the lightness of the masked sealant area. A sealant showing no dirt accumulation and no change in color has a dirt collection index ($D_c$) of 100.

Table 3 shows the original sealant color as well as the dirt collection indices measured after 11 and 23 months of exposure. Results obtained after 11 months of weathering are reproduced from [2]. Fig. 21 shows the specimens after 27 months exposure. As can be seen, the two polyurethane sealants (20-11 and 20-12) as well as the two-part surface modified silicone sealant (20-8) show the lowest dirt pick-up ($D_c$>95) after 23 month exposure. The dirt collection indices of these sealants also changed little ($\Delta D_c$<2) between 11 and 23 month exposure. The one-part polyurethane sealant (20-12) did, however, receive a dirt collection index above 100, due to the chalking that caused the surface to appear lighter than the masked portion. Interestingly, the one-part surface modified silicone sealant (20-7), showing no fluid migration, and the two highly plasticized silicone sealants (20-5 and 20-6), which were the worst sealants from a fluid migration point of view, received similar dirt collection index ratings and also showed comparably small changes in dirt pick-up ($\Delta D_c$<2) between 11 and 23 month exposure. One of the silicon modified polyether sealants (20-10) and two plasticized silicone sealants (20-1 and 20-4) picked up moderate amounts of dirt ($\Delta D_c$<4) between 11 and 23 month exposure and show dirt collection indices in the middle range. The other silicon modified polyether sealant (20-9) and, interestingly, the unplasticized silicone sealant (20-2) picked up considerable amounts of dirt ($\Delta D_c$>9) between the 11 and 23 month exposure and show dirt collection indices at the end of the scale. Interestingly, none of the sealants oriented at a 90° angle to horizontal showed signs of mildew growth, which led to increased dirt pick-up on their 45° angle weathered counterparts.

TABLE 3 -- Dirt collection ratings $D_c$ according to ASTM D3719 after 11 and 23 month exposure in vertical position at Northern, industrial site

| Sealant | Type[1)] | Original Color | $D_c$ Index | |
|---|---|---|---|---|
| | | | 11 Months | 23 Months |
| 20-1 | Si/P/1 | white | 89.15 | 86.41 |
| 20-3 | Si/P/1 | off-white | 92.5 | -----[2)] |
| 20-4 | Si/P/1 | off-white | 87.7 | 84.51 |
| 20-5 | Si/P/1 | off-white | 91.36 | 91.71 |
| 20-6 | Si/P/1 | grey | 93.91 | 92.23 |
| 20-2 | Si/U/1 | off-white | 89.25 | 78.44 |
| 20-7 | Si/U/SM/1 | off-white | 92.63 | 92.98 |
| 20-8 | Si/U/SM/2 | off-white | 94.75 | 95.05 |
| 20-9 | MPe/P/1 | white | 89.62 | 80.23 |
| 20-10 | MPe/P/1 | tan | 93.29 | 89.35 |
| 20-11 | Pu/P/2 | limestone | 96 | 94.88 |
| 20-12 | Pu/U/1 | white | 100.21 | 100.95 |

1) Sealant types same as in Table 1
2) Sample 20-3 was lost after 11 months exposure due to weather conditions

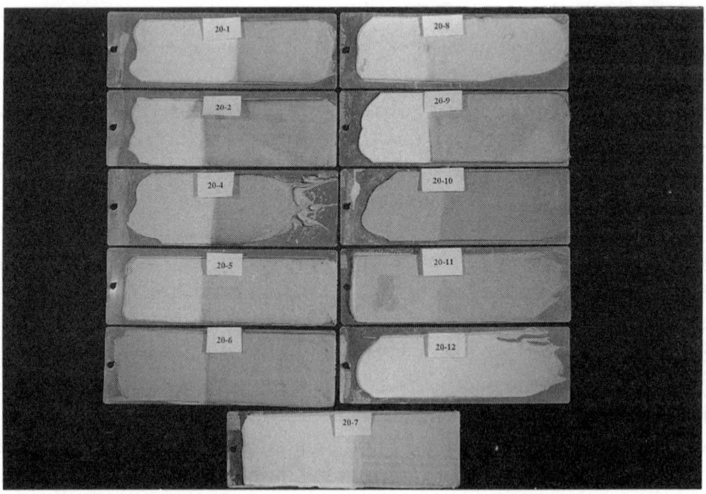


FIG. 21 Dirt pick-up on vertically exposed sealant samples after 27 months exposure at Northern, industrial site.

Based on the dirt collection indices measured after 23 months outdoor exposure of sealant samples in vertical position at the Northern, industrial exposure site, the sealants can be ranked with regard to their dirt pick-up as follows:

20-12 (Pu/U/1) < 20-8 (Si/U/SM/2) ≤ 20-11 (Pu/P/2) < 20-7 (Si/U/SM/1)
≤ 20-6 (Si/P/1) < 20-5 (Si/P/1) < 20-10 (MPe/P/1) < 20-1 (Si/P/1)
< 20-4 (Si/P/1) < 20-9 (MPe/P/1) < 20-2 (Si/U/1)

As can be seen by comparing Tables 1, 2 and 3, there is a significant difference in dirt pick-up ratings between vertically and inclined exposed samples. Dirt pick-up, thus, not only depends on the nature of airborne particulates present at the exposure site, but also on the specific exposure conditions.

## CONCLUSIONS

Some general observations include that the environment and exposure conditions play a significant role in the amount of fluid streaking, dirt pick-up and mold growth exhibited by a sealant as seen by the differences in results between exposure of test sealants in industrial Michigan versus rural Georgia. Fluid migration, on the other hand, is relatively unaffected by the environment under the test conditions chosen for this study (static joints). The most noticeable differences are the rate of fluid migration and, for sealants containing degradable organic plasticizers, the relative rates of degradation and replenishment, which are affected by the environment. The outdoor weathering results also show that for some sealants fluid migration may not be noticeable for almost a full year. It is questionable, whether the accelerated ASTM C1248 test would be capable of predicting this delayed staining caused by unreacted polymers.

Within 18 months of inclined weathering at the Northern, industrial site, the plasticized and unplasticized silicone sealants show significant to severe levels of fluid streaking and dirt pick-up. However, a vast range of fluid migration into marble still exists. The two unplasticized, surface-modified silicones (20-7 and 20-8) and the polyurethane sealants (20-11 and 20-12) show no fluid migration into marble and only slight fluid streaking on reflective glass after 25 months of outdoor weathering. The silicon modified polyethers (20-9 and 20-10) show significant levels of fluid migration within the first 18 months of weathering, with the duration of the staining depending on the formulation. Mold growth on the inclined samples of the polyurethanes (20-11 and 20-12) and the two-part surface modified silicone (20-8) prevents accurate comparisons of dirt pick-up. Dirt pick-up for the vertically exposed samples, however, proves to be independent of polymer type.

## SUMMARY

The general finding of this paper is that there are factors other than the chemical nature of the polymeric backbone that influence the aesthetic performance of weatherproofing sealants: First, the extent and rate of fluid migration into porous substrates mainly depend on the viscosities and quantities of free polymers and fluids in a sealant. Substrate porosity and color are also key factors in the extent and visibility of fluid migration. Second, dirt pick-up during cure is primarily influenced by the speed of surface cure and the conditions of the environment. Three mechanisms for dirt pick-up on a cured sealant surface were proposed: a) residual surface tack leading to imbedded dirt particulates; b) dirt pick-up resulting from migration of high(er) molecular weight species to the sealant surface; and c) physical dirt adsorption which is primarily related to the viscoelastic properties of the sealant's elastomeric network. Dirt retention is influenced by the durability of the sealant's surface. Third, due to its dependence on migrating fluids, fluid streaking is also primarily a formulation dependent phenomenon which can occur for any sealant type. Organic sealants, if formulated with degradable plasticizers, tend to show temporary or cyclic fluid streaking, while silicone sealants, due to the environmental stability of free fluids contained in them, tend to show more permanent fluid streaking. Lastly, chalking refers to the whitening of some organic sealant surfaces with weathering and is caused by the degradation and erosion of the polymeric matrix leading to increased roughness or porosity of the sealant surface. In addition to lightening of the original sealant color, "white run-down" and mold fungus growth may occur. In cases where mold is not substantially present, chalking can act as a cleansing mechanism to remove sealant dirt pick-up. However, a sealant that is lighter in color, versus darker due to dirt pick-up, still presents an aesthetic issue.

While the ASTM C1248 test method is useful for sealant manufacturers, who know the formulation of the sealant they study, it can be misleading, if used by someone not having this information. High molecular weight plasticizers and unreacted polymers, which take considerable time to migrate from the cured sealant, are not detected by this test method. For this reason, comparative outdoor exposure tests for dirt pick-up, fluid migration and fluid streaking are proposed. Chalking of organic sealants can be accelerated by exposing them to UV light, heat and humidity in a weatherometer for at least 2000 hours.

**REFERENCES:**

[1] Kawamura, M. and Kuwabara Y., "Aging of Facades", Nikkei Architecture, Nov. 1994, pp 82-111.

[2] O'Neil, V., Klosowski, J., Altes, M., and Bergman, L., "Effects of Weatherproofing Sealants on Building Aesthetics", Science and Technology of Building Seals, Sealants, Glazing and Waterproofing - Fourth Volume, ASTM STP 1243 David Nicastro, Ed., American Society for Testing and Materials, Philadelphia, 1994.

[3] Bridgewater, T.J., and Carbary L.D. "Accelerated Weathering and Heat Stability of Various Perimeter Sealants", Science and Technology of Building Seals, Sealants,

Glazing and Waterproofing: Second Volume, ASTM STP 1200, Jerome M. Klosowski, ed., American Society for Testing and Materials, Philadelphia, 1992, pp. 10-28.

[4] Klosowski, J.M., Sealants in Construction, Marcel Dekker, Inc., New York and Basel, 1989, pp 200-212.

[5] Gvozdic, N.V., "Origin and Mechanism of Substrate Staining", Dow Corning Internal Research Report, April 1990.

[6] Krupp, H., "Particle Adhesion. Theory and Experiment", Advances in Colloid and Interface Science, Vol. 1, May 1967, pp. 111-239.

[7] Das, S.K., Schechter, R.S., and Sharma, M.M., "The Role of Surface Roughness and Contact Deformation on the Hydrodynamic Detachment of Particles from Surfaces", Journal of Colloid and Interface Science, Vol. 164, 1994, pp. 63-77.

[8] Friedman, M., Casati, F., and Stamm, J., "Selecting the Proper Biocide for Adhesives and Sealants", Adhesives Age, February 1988.

[9] Gladstone, M.E. and Yarosh, K.F., unpublished results of field inspections, 1990-1995.

[10] Wolf, A.T., unpublished results on outdoor weathering of various polysulfide and polyurethane sealant formulations, 1981.

[11] ASTM C1248 "Standard Test Method for Staining of Porous Substances by Joint Sealants", American Society for Testing and Materials, Philadelphia, 1993.

[12] ASTM D2203 "Standard Test Method for Staining from Sealants", American Society for Testing and Materials, Philadelphia, 1993.

[13] Norme Francais NFP 85-512 "Mastic du Type Plastique Utilises Pour le Calfeutrement Etanche des Joints - Essai de Diffusion des Constituants", Association Francaise de Normalisation (AFNOR), Paris, 1985.

[14] British Standard BS 3712 "Methods of Test for Building Sealants. Part 2. Seepage, Staining, Shrinkage, Alkali Resistance, Shelf Life, and Paintability", British Standards Institution (BSI), London, 1973 and 1985.

[15] Gvozdic, N.V., "Accelerated Staining and Dirt Pick-Up Test", Dow Corning Internal Research Report, April 1990.

[16] Leempoel, P. and Chasseur, A., "Hybrid Sealants - Part II", Dow Corning Internal Research Report, January 1995.

[17] Fedor, G.R., "Usefulness of Accelerated Test Methods for Sealant Weathering", Science and Technology of Building Seals, Sealants, Glazing and Waterproofing: Second Volume, ASTM STP 1200, Jerome M. Klosowski, ed., American Society for Testing and Materials, Philadelphia, 1992, pp. 48-50.

[18] Dirt pick-up and fluid streaking ratings were adapted from those of US Patent 4,695,603, pp. 8, lines 41-52.

[19] ASTM D3719 "Standard Test Method for Dirt Collection Index", American Society for Testing and Materials, Philadelphia, 1987.

Ian R. Chin,[1] Todd A. Gorrell,[2] and Michael J. Scheffler[3]

# STAINING POTENTIAL OF SEALANTS IN/ON EXTERIOR WALL SUBSTRATES

**REFERENCE:** Ian R. Chin, Todd A. Gorrell, and Michael J. Scheffler **"Staining Potential of Sealants In/On Exterior Wall Substrates,"** Science and Technology of Building Seals and Sealant: Fifth Volume, ASTM STP 1271, Michael A. Lacasse, Ed., American Society for Testing and Materials, 1996.

**ABSTRACT**: The investigation of dozens of exterior walls on buildings by the authors have revealed that certain sealants used to seal joints in exterior walls have caused the following conditions:

1. Local staining and discoloration of masonry wall substrates due to penetration of sealant into substrates.
2. Local change in absorption of masonry wall substrates due to penetration of sealant into the substrates which results in local discoloration of wall substrates after rains.
3. Surface discoloration of sealant due to dirt accumulation on sealant.
4. Stains in the form of streaks of dirt on masonry and on metal wall substrates due to transfer by rain water of some of the dirt on the sealant surface onto the adjacent surface of the substrate.

Laboratory and field testing of the major commercially available sealants by the authors to date have revealed that silicone sealants have the propensity to stain certain exterior wall substrates while polyurethane sealants do not; that silicone sealants in service discolor more significantly and faster due to dirt accumulation than polyurethane sealants; that sealant that has penetrated into masonry substrates cannot be entirely removed from the substrate by cleaning; and that preconstruction testing of sealants and proper material selection can help to avoid staining of sealant.

**KEYWORDS:** ASTM, cleaning, dirt, oil penetration, polyurethane, poultice, sealant, silicone, stain, substrate, testing

[1]Vice President and Principal, Wiss, Janney, Elstner Associates, Inc., Chicago, IL

[2]Senior Architect/Engineer, Wiss, Janney, Elstner Associates, Inc., Chicago, IL

[3]Consultant, Wiss, Janney, Elstner Associates, Inc., Chicago, IL

## INTRODUCTION

Staining of exterior wall substrates caused by sealants can adversely affect the appearance and image of buildings. The staining potential of wall material substrates due to sealants is known within the sealant industry. The American Society for Testing and Materials (ASTM) tests to evaluate the staining potential of sealants are ASTM "Standard Test Method for Staining and Color Change of Single-or Multicomponent Joint Sealants" (ASTM C 510), [1]; ASTM "Standard Test Method for Staining of Pourous Substrates by Joint Sealants" (ASTM C 1248), [2]; and ASTM "Standard Test Methods for Staining from Sealants" (ASTM D 2203), [3]. However, when the ASTM C510 and ASTM D 2203 tests are performed prior to the selection of the wall substrate they may not predict the potential of a sealant to stain a specific wall substrate, because these tests utilize filter paper and mortar to represent the wall substrate and these test materials are not similar to masonry (stone, brick, concrete block) wall substrates.

ASTM C 1248 utilizes substrates such as stone and is an accelerated test that lasts up to 28 days. ASTM C 1248 states "since this is an accelerated test, it does not necessarily predict that the tested sealants will not stain or discolor pourous substrates over longer periods of time."

In addition, these tests do not address sealant stains resulting from dirt accumulation and dirt runoff. Also, many building owners, designers, and wall material manufacturers and suppliers are unaware of the staining potential of some sealants in contact with materials used in many exterior walls. Understanding of the various types of stains and of the conditions under which stains occur can help owners, designers, and wall material manufacturers and suppliers minimize the occurrence of this problem on their buildings.

## BACKGROUND

There are many types of sealant in use today, but the two predominant types of high performance sealant currently used in the majority of sealant applications in exterior walls on buildings are silicone and polyurethane. A high performance sealant is a sealant that can extend and compress 25 percent or more of its width without detrimental effects. Prior to the 1980's, silicone and polyurethane sealants generally had their own niche in the construction industry. Silicone sealants were generally used in the construction of aluminum and glass curtain wall systems, while polyurethane sealants were generally used in masonry (stone, brick, concrete block) exterior wall systems. This general usage of sealant occurred because the thermal expansion of aluminum is about three times that of masonry, and silicone sealant could extend and compress twice that of polyurethane sealant manufactured at that time. For substrates such as masonry and concrete where joint movement was typically less of a concern, less expensive polyurethane sealant and occasionally, polysulfide sealant were used.

However, in the 1980's, this general usage of sealants changed because construction of new buildings occurred worldwide in large numbers, and designers moved away from aluminum and glass curtain walls of the previous decades towards masonry and concrete cladding. Also, the material cost difference between silicone and polyurethane sealants became less of a factor because of increasing labor costs associated with construction. The designers of the new buildings also began to require more extension and compression capabilities from sealants because of the desire to have fewer and narrower sealant joints for aesthetic reasons. Due to these new trends in the building industry, sealant manufacturers began to compete with each other more heavily outside their established areas of general usage. Consequently, the following occurred:

1. The use of silicone sealants to seal joints in masonry and concrete walls increased.
2. Staining of some masonry exterior wall substrates where silicone sealant had been used became more evident.
3. Polyurethane sealant manufacturers developed sealant with more extension and compression capability than that previously available.
4. Polyurethane sealant manufacturers developed a new polyurethane sealant that cures faster than the original polyurethane sealant.
5. Silicone sealant manufacturers began to develop a non-staining, non-dirt pickup silicone sealant.

## TYPES OF SEALANT STAINING

The types of stains that occur on wall substrates due to sealant are as follows:

1. Oil Penetration Stain--Staining and change in appearance of wall substrate directly adjacent to the sealed joint caused by the penetration of oils from the sealant into a masonry substrate, as shown in Fig. 1.
2. Halo Stain--Change in absorption of wall substrate directly adjacent to the sealant joint at areas of sealant oil penetration. These areas become hydrophobic and do not absorb water at the same rate as the unaffected remaining area of the panel. This condition results in an apparent perimeter "halo" when the panel is wet, as shown in Fig. 2.
3. Joint Surface/Dirt Accumulation Stain--Accumulation of dirt and debris and other airborne particles on the surface of the sealant joints, as shown in Fig. 3.
4. Surface Dirt Runoff Stain--Staining of a masonry or metal substrate surface caused by runoff of dirt that has accumulated on the surface of the sealant joint, as shown in Fig. 4.

Investigation of stained masonry substrates, by the authors has revealed that the above types of stains were primarily caused by silicone sealants. Staining of masonry substrates by silicone sealants have been previously documented by others.[3],[4],[5]

FIG. 1--View of sealant oil penetration stain into masonry substrate.

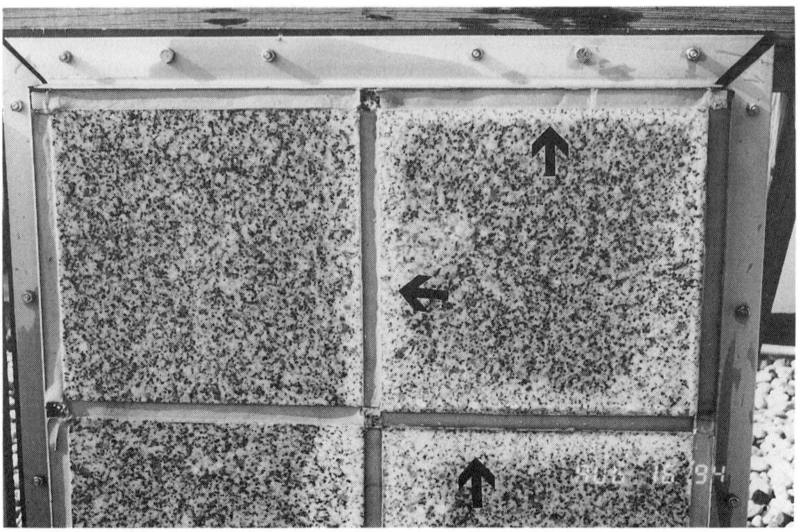

FIG. 2--View of halo stain.

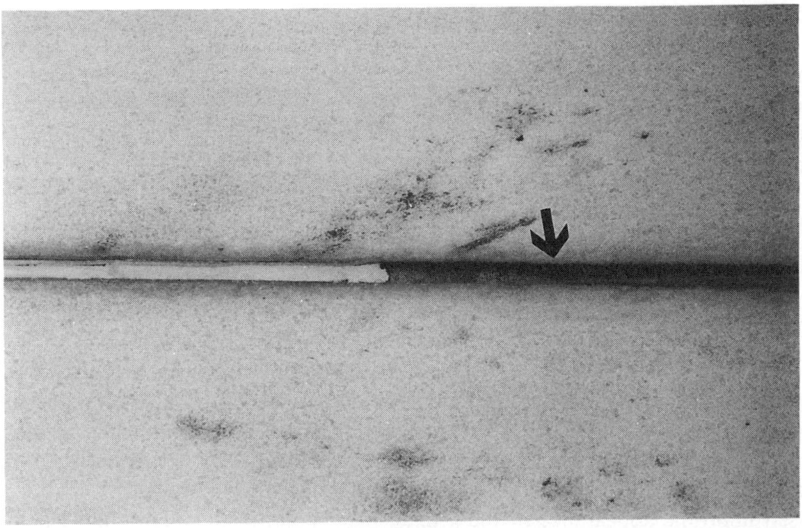

FIG. 3--View of joint surface/dirt accumulation stain.

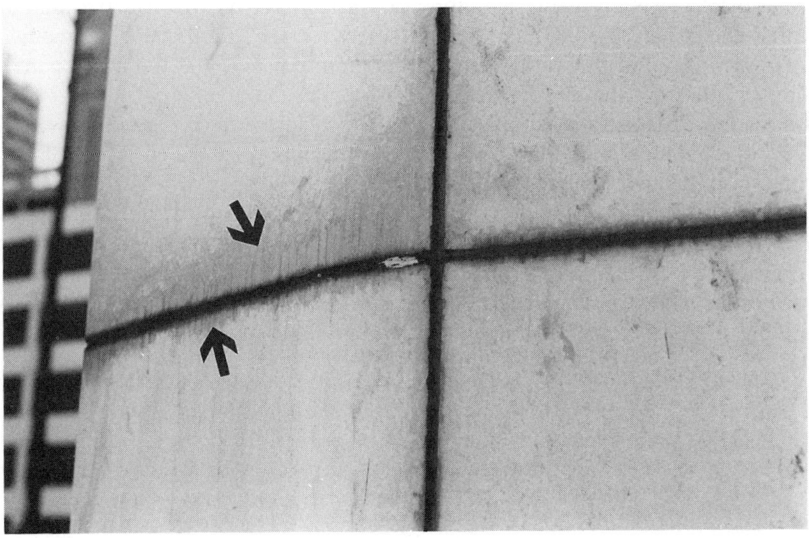

FIG. 4--View of surface dirt runoff stain.

## CASE STUDY: THE AMOCO BUILDING

The Amoco Building in Chicago, Illinois, is an 82 story structure that was constructed in 1972. The building was clad with approximately 44,000 white marble panels that were approximately 1.22m (4 ft) tall by 1.22m (4 ft) wide by 32mm (1-1/4 in.) thick.

In 1988, the decision was made by Amoco Properties, Inc. (Amoco) to remove the existing white marble panels from the building due to bowing, cracking, and deterioration of the marble panels. The authors were a part of the team assembled by Amoco to design the new cladding for the building. In order to aesthetically evaluate potential cladding materials to replace the original marble panels on the building, a full-size mock-up of various cladding materials was constructed at Amoco's oil refinery site in Whiting, Indiana. Two types of white granite panels and two types of white metal panels were installed and evaluated on the mock-up. The joints between the granite panels on the mock-up were sealed with polyurethane sealant, and the joints between the metal panels were sealed with silicone sealant. The joints between granite panels and metal panels were sealed with silicone sealant.

Based upon the appearance of the granite panels and of the aluminum panels on the completed mock-up, and upon the building having been previously clad with a white stone, Amoco decided to reclad the building with white granite panels to maintain its original appearance and image.

However, in the dusty environment of the oil refinery site where the mock-up was constructed, the joints between the metal panels that were sealed with silicone sealant accumulated dirt very quickly, and within a few days the surface of the silicone sealant, that were originally white in color, had turned dark brown. During this time, the polyurethane sealant used to seal the joints between the granite panels remained relatively clean. Also, at the joints between the granite panels and the metal panels, penetration of oils from the silicone sealant had stained the granite panels.

As a result of the sealant staining that developed in the mock-up, a test was designed by the authors to determine the staining potential of different types of silicone and polyurethane sealants in contact with the two types of white granite being considered for the building. The test utilized a frame consisting of four, 305mm (12 in.) square white granite panels, with 13mm (1/2 in.) wide sealant joints. Six sealants were installed between the granite panels in the test frame, two silicones and four polyurethanes, as shown in Fig. 5. The test panel was placed on the roof of a two-story building in Northbrook, IL, facing south at an angle of 30 degrees from the horizontal.

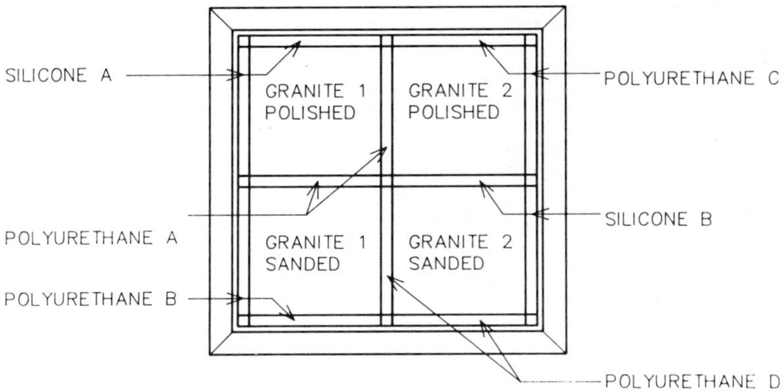


FIG. 5--Schematic elevation of test panel.

After approximately 6 months of exterior exposure, the surfaces of the two silicone sealants in the test frame had darkened in color due to dirt accumulation, and minor penetration of oils into the granite panels of approximately 3mm (1/8 in.) from the silicone sealants had occurred. After 1 year and 6 months of exterior exposure, extensive penetration of oils from the silicone sealants into the granite panels of approximately 25mm (1 in.) to 75mm (3 in.) and dirt accumulation and dirt runoff from the silicone sealants had stained the granite panels. Fig. 6 is a view of the test panel when it was placed on the building on September 15, 1988, and Fig. 7 is a view of the test panel approximately 18 months later, on March 1, 1990. The polyurethane sealants had not stained the granite panels and were relatively free of dirt accumulation.

FIG. 6-- A view of test panel when it was placed on the roof of the building on September 15, 1988.

FIG. 7--View of test panel after approximately 18 months of exposure, on March 1, 1990.

As a result of this testing, a two-part polyurethane sealant was preliminarily selected to be used to seal the joints between the new granite panels on the Amoco Building.

This preliminary selection of sealant was discussed with Amoco maintenance personnel that were going to be responsible for the maintenance of the new granite cladding on the building. The maintenance personnel stated that they would prefer to use a one-part silicone sealant on the building because silicone sealants are normally premixed and are available in tubes and are therefore easier to use and are less wasteful than a two part-polyurethane sealant.

Based on the concerns of the Amoco maintenance personnel, Amoco decided to test and evaluate on the building itself, the performance of available silicone and polyurethane sealants installed in joints between full-size white Mt. Airy granite panels that were selected for recladding the building. The manufacturer of every major commercially available sealant was invited by Amoco to participate in the test. Each sealant manufacturer that agreed to participate in the test was told by Amoco that if their sealant stained the granite it would not be selected for use on the building. They were, therefore, asked to carefully study the substrate material to determine which of their sealants was best suited for the project, and which would not stain the stone, and to determine if a primer was required. Each sealant manufacturer was also asked to clean the substrate to their specifications and to use a contractor of their choice to install the sealant in their area of the test.

In response to this request, five sealants from five different manufacturers were tested on the south elevation at the third floor of the building. Each of the five test areas consisted of four full-size granite panels with sealant in the joints between panels. The five sealants tested included two polyurethanes and three silicones.

A similar test was conducted in the laboratory by the authors using the same sealants and the same granite to confirm the results of the test on the building. The sealants in the laboratory test were installed by an experienced sealant contractor in accordance with the sealant manufacturer's recommendations in test frames made with four, 305mm (12 in.) square panels of granite.

The test areas on the building and the laboratory test were monitored regularly. Within six to nine months of the installation of the sealant in the test areas, oil penetration and halo staining of the granite panels adjacent to joints sealed with the three silicone sealants occurred. The surfaces of the silicone sealants had accumulated dirt to varying degrees, with some exhibiting heavy dirt accumulation. Also, when the panels were sprayed with water, all of the panels sealed with silicone sealant were observed to display halo staining that was approximately 25mm (1 in.) to 38mm (1-1/2 in.) wide on the building panels, and that was approximately 25 mm (1mm) to 75mm (3 in.) wide on the laboratory test panels adjacent to the sealant joints, confirming that oils from the silicone sealant had penetrated into the granite, as shown in Figs. 8 through 11. The two polyurethane sealants in the test did not stain the granite panels.

Based upon the results of all of the above testing, Amoco decided to use a two-part polyurethane sealant to seal joints between the new Mt. Airy granite panels on the building.

FIG. 8-- View of four of the five test areas on the Amoco building. The fifth test area, sealed with a polyurethane sealant, is not shown. Test area Nos. 1, 3, and 4, were sealed with silicone sealant. Test area Nos. 2 and 5 (not shown) were sealed with polyurethane sealant.

FIG. 9-- View of 4 of the 5 test areas on the Amoco building after approximately 6 to 9 months of exterior exposure showing halo stains adjacent to joints sealed with silicone sealant.

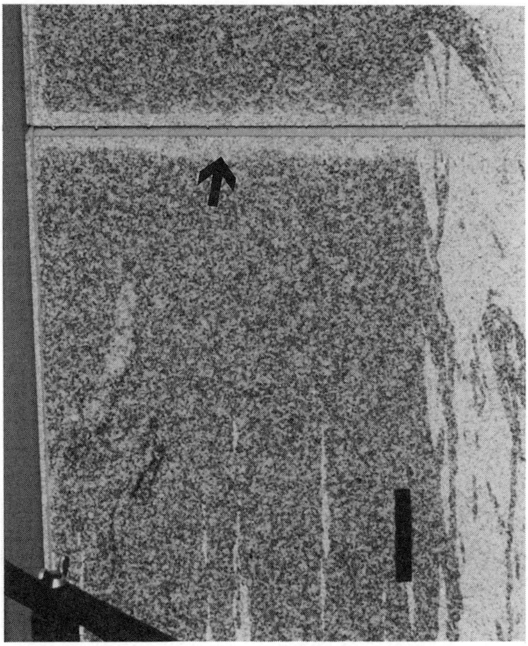

FIG. 10--Close-up view of halo stain at joint sealed with silicone sealant on test area No. 1.

FIG. 11-- View of halo stains adjacent to joints sealed with silicone sealants in test panels constructed by authors.

## CLEANING OF SEALANT STAINING

Once staining of masonry due to penetration of oils from sealants has occurred, attempts can be made to remove the stains. However, prior to beginning any cleaning, the sealant that caused the staining should be completely removed from the joints. If this sealant is not removed from the joints, oil penetration, and halo staining will most likely continue.

The most common types of cleaning methods that are available to clean masonry wall substrates that have been stained by sealant include the following:

1. Water Washing--Water washing includes such methods as bucket and brush cleaning; low and high pressure water sprays. Washing with water is typically a mild and safe method of cleaning. Water washing may be effective in removing dirt accumulation and dirt runoff on the surface of masonry. The water wash may be modified to include mild detergents to help remove surface oils. However, oils that have penetrated into the masonry cannot be removed by water washing and will continue to stain the masonry.

2. Mechanical Cleaning--Mechanical cleaning includes brushing; scraping; and abrasive blasting, such as sand blasting. While mechanical cleaning may be effective in removing surface dirt accumulation and dirt runoff, it cannot remove oils that have penetrated into the masonry. Mechanical cleaning may also damage the masonry substrate and is not recommended without a thorough evaluation as to its effect on the substrate.

3. Chemical Cleaning--Chemical cleaning usually involves diluted solutions of alkalis or acids, such as sodium hydroxide or hydrofluoric acid. These solutions can be effective in removing surface dirt but are not effective for removing oils that have penetrated into the masonry. Chemical cleaning may also damage the masonry substrate and is potentially dangerous to the workers if not used properly.

4. Poultice Cleaning--Poultices made of chemical agents such as ethylene diamine-tetra-acetic acid are mixed with a medium such as cellulose or diatomaceous earth to form a thick paste or gel. Poultice cleaning consists of applying the poultice onto the stained areas of masonry and allowing the poultice to dry with the intent of pulling out oils that have penetrated into the substrate. Usually several applications of the poultice are required, but even then, this cleaning method is rarely completely effective in the removal of oils from silicone sealants that have penetrated into masonry substrates.

The above types of cleaning methods have not been effective in the total removal of stains in masonry wall substrates caused by the penetration of oils from sealants. If these oils are not completely removed from the substrate, dirt accumulation will usually continue to occur even after the above cleaning methods have been used, resulting in the need for additional periodic cleaning. Preconstruction investigation and testing to select a sealant that will not stain the selected wall substrate is, therefore, essential.

## PREVENTION OF STAINING

The most effective way to avoid unsightly staining from sealants is to prevent the staining from occurring by selecting a sealant that will not stain the substrate. Preconstruction testing of sealants with the actual materials to be used in the construction should be performed. Normally, preconstruction testing to determine the staining potential of sealants on substrates is performed by the sealant manufacturer. The test method used is either that specified by ASTM C 510 [1], ASTM D 2203[2], ASTM C 1248, [3], or a test method that the sealant manufacturer has developed.

ASTM C 510 and ASTM D 2203 evaluate the likelihood of a sealant causing early stain on a porous substrate due to exudation from the sealant. ASTM C 510 utilizes test specimens consisting "of a slab of mortar upon which is placed a layer of sealant." ASTM D 2203 utilizes a test specimen consisting of 10 filter papers stapled together upon which a layer of sealant is placed. ASTM D 2203 has an alternate test procedure that utilizes a test specimen consisting of a sample of the wall substrate upon which a layer of sealant is placed. However, the sample is only exposed to this test for 5 days. Therefore, a pre-construction claim made prior to selection of the wall substrate stating that a sealant meets the ASTM C 510 and the ASTM D 2203 stain test, does mean that post construction staining will not occur on masonry wall substrates, because the properties of the filter paper and mortar ASTM test specimens are not the same as masonry substrates.

ASTM C 1248 is an accelerated laboratory test to determine if a sealant has a probability of staining a pourous substrate such as stone. This test may be more appropriate than ASTM C 510 and D 2203 to evaluate staining potential of sealant on stone and masonry substrates. However, "it does not necessarily predict that the tested sealant will not stain or discolor pourous substrates over longer periods of time". [3]

Some sealant manufacturer's have developed their own test methods to evaluate staining potential of their sealant on wall substrates. Normally, these test methods are not published. If a sealant manufacturer proposes to use his own test method, a copy of the test method should be obtained and reviewed to determine if it is appropriate for the specific project.

Another test method that may be used is timely, real-time testing using test frames constructed with the wall substrate materials to be used on the project, as described earlier for the Amoco Building. This test frame should be exposed to the same environmental conditions as the project. Samples of potential sealants to be used for the job should be installed in the test frames by an experienced sealant installer in accordance with the sealant manufacturer's written instructions. The test frames should be regularly monitored, in both wet and dry conditions, and the conditions of the sealant and substrate should be recorded.

The real-time test frame is the most accurate and reliable test method presently available. However, this method may require a lead time of several months to adequately assess the staining potential of the sealants used in the test.

## CONCLUSIONS

In the last 15 years, the occurrence of staining of masonry substrates caused by sealants has increased. Investigations and testing by the authors have revealed that sealant staining of masonry substrates is usually caused by silicone sealants. Staining is not usually caused by polyurethane sealants.

When staining has occurred on a building, cleaning of the substrate to try to remove the stains can be performed. However, cleaning is not usually fully effective in removing all traces of the stain, especially stains caused by penetration of oils from the sealant into the substrate.

Prevention of staining, and determination of the staining potential of sealants with specific substrates can be achieved with a proper preconstruction testing program. Testing should be performed for every project where the potential of staining of the wall substrate exists, because sealant formulations could be changed at any time without notification to consumers and, therefore, a sealant that was previously tested with a specific substrate material with no signs of staining may not be free from risk of staining in the future.

Current ASTM test methods performed before the wall substrate is known may not determine the staining potential of sealants on a specific substrate. Test methods that have been developed by sealant manufacturers should be reviewed carefully to determine whether or not they are appropriate for the project.

Real-time testing of sealants with the selected substrate materials and the identical environmental conditions of the project is the most reliable test method currently available, and should be performed. However, this method may require up to a year for reliable results.

The authors are currently performing research with the intent of developing a test method to reliably determine the staining potential of sealants on masonry substrates in a shorter period of time. When this test method is developed it will be made available to the industry.

**REFERENCES**

[1] American Society for Testing and Materials, ASTM C 510 "Standard Test Method for Staining and Color Change of Single-or Multicomponent Joint Sealants," Annual Book of ASTM Standards, Vol. 04.07, pp 17, Philadelphia, PA, 1993.

[2] American Society for Testing and Materials, ASTM D 2203 "Standard Test Methods for Staining From Sealants," Annual Book of ASTM Standards, Vol. 04.07, pp 256, Philadelphia, PA, 1993.

[3] American Society for Testing and Materials, ASTM C 1248 "Standard Test Methods for Staining of Pourous Substrates by Joint Sealants," Annual Book of ASTM Standards, Vol. 04.07, pp. 308, Philadelphia, PA.

[4] Chipman, Jerry "Contractor Challenges: Staining, Rundown," Adhesives Age, January, 1987.

[5] Honold, Richard "Staining Caused by the External Use of Silicone Rubber Sealants", Bau Gewerbe, Vol. 11/82, Rudolf Muller GmbH, Cologne, F.R. Germany.

[6] Panek, Julian R. and Cook, John Philip, Construction Sealants and Adhesives, 3rd ED., pp 76.

Matthew C. Farmer[1], and Richard A. Cechner[2]

# LABORATORY TESTING OF SEALANTS WITH A MARBLE SUBSTRATE

**REFERENCE:** Farmer, M.C., and Cechner, R.A., **"Laboratory Testing of Sealants with a Marble Substrate,"** Science and Technology of Building Seals, Sealants, Glazing and Waterproofing: Fifth Volume, ASTM STP 1271, Michael A. Lacasse, Ed., American Society for Testing and Materials, 1996.

**ABSTRACT:** Polyurethane and silicone sealants are widely specified for installations with porous substrates such as some stones. However, when a sealant is used against such surfaces, there is a potential for a lack of adequate adhesion, or staining of the stone by misapplication or migration of the liquid components of the sealant system, such as primers or plasticizers in the formulation. Some varieties of marble in particular have been reported to be susceptible to staining and discoloration over time from sealants. Application of new sealant over existing sealants is also of great concern for remedial applications. Determining the level of substrate preparation necessary to achieve adequate bond is critical to the success of the remedial construction project.

This paper discusses the development and results of a test program conducted to determine the relative performance of sealants installed on a white marble substrate. The tests performed included wet adhesion tests, accelerated weathering studies, and staining due to plasticizer migration.

**KEYWORDS:** Sealants, Marble, Testing, Adhesion, Accelerated Weathering, Staining

Elastomeric joint sealants are one of the most common building materials in use today. They are used every day in the construction industry as a primary defense against moisture intrusion. However, sealants are also extremely vulnerable to conditions that can reduce their durability and performance over the long term, particularly their adhesion characteristics. If improperly selected or applied, the results can be disastrous, leading to degradation of other building

[1] Senior Engineer, Washington D.C. office of Wiss, Janney Elstner Associates, Inc., 2750 Prosperity Avenue, Suite 200, Fairfax, VA 22031, USA.

[2] Senior Chemist, Erlin Hime Associates Division of Wiss, Janney, Elstner Associates, Inc., 330 Pfingsten Road, Northbrook, IL 60062, USA.

materials, water damage and, at the very least, premature failure of the sealant material. The challenge to the design professional is to discover potential problems in the laboratory, not on the job site.

In order to understand the behavior of a particular sealant in an intended application, it can be useful to simulate conditions under which the sealant is to be used and expected to perform. This is best achieved through a combination of field and laboratory testing. Testing can be designed to simulate the worst conditions that the sealant is likely to undergo during its service life. By performing specially designed tests in the laboratory, one can determine the relative performance of a sealant on a particular substrate more quickly than evaluating a field application and waiting for extreme conditions to occur.

This paper presents an example of a testing program for replacement sealant selection for a marble substrate. It reviews and discusses the program structure and results in order to demonstrate the effectiveness of testing and the knowledge to be gained from a laboratory testing program.

This testing program examined the key issues of sealant adhesion, compatibility, and long-term performance, as well as joint surface preparation and staining. This testing program contrasted four sealants, comparing their performance in the aforementioned areas. Although field test installations were also conducted to evaluate adhesion and compatibility, this paper addresses only the laboratory portion of the testing program. The results of the testing are presented to illustrate how such a program can be used to compare objectively the performance of several sealants, and to determine the most appropriate sealant for a particular application. The findings are not intended to provide specific information about the performance of any particular sealant or sealant product.

Porous stone substrates such as marble are among the most difficult substrates to seal successfully for the long term. The absorption and retention of water within the substrate around the joint sealant can undermine any joint installation by disrupting the sealant bond. Porous stone is also more likely to absorb fluid components of the sealant that can cause visible and objectionable staining, particularly in stone such as light-colored marble. Reseal applications can be especially difficult due to existing, potentially incompatible sealants or their components, being absorbed and remaining within the stone substrate, thus interfering with the adhesive qualities of the new sealant.

## PROJECT DESCRIPTION

The subject project for this sealant testing program was an existing building, with an exterior wall system consisting of white marble-faced precast panels. The structure is approximately 25 years old. The original sealant had not been replaced or maintained and was severely deteriorated. Sealant deterioration was observed throughout the wall system in the form of widespread missing and unbonded sections of sealant, a lack of flexibility, and cracking within the sealant bead (Fig. 1). Studies indicated that the joints had failed due to a combination of age, weathering, and water infiltration behind the sealant, causing a deterioration of the original sealant polymer[3]. Some of the joints, which ranged in width from 3 mm to 13 mm (1/8 in. to 1/2 in.), undergo limited panel-to-panel movements as the building is subjected to service lateral loads.

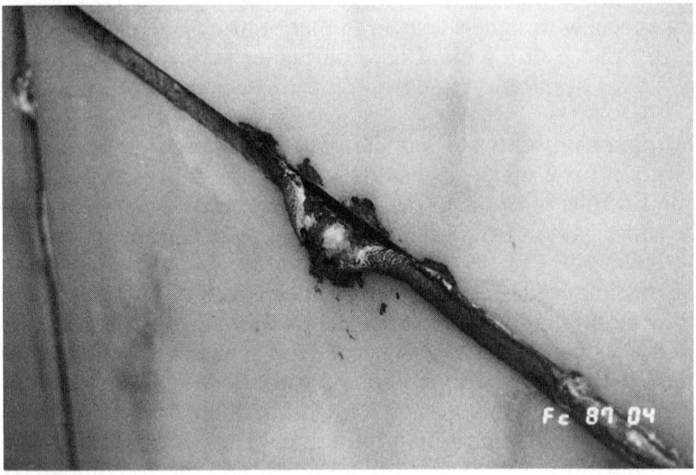


FIG. 1 -- Typical original joint sealant condition. Note breach caused by water migrating out from behind the sealant.

## SEALANT TESTING PROGRAM OBJECTIVES

The objective of this sealant testing program was simple: select and compare several potential sealants to determine which could be expected to perform most effectively given the existing substrate and environmental conditions. Specifically, the test program was developed to evaluate the following issues critical to a successful sealant installation in a white marble substrate:

---

[3]The original selection of a relatively high modulus, low elongation type of sealant (an acrylic based polymer) may have also contributed to the failure in joints with excessive movements.

Adhesion to substrate--Adhesion is one of the most important characteristics of any sealant installation. It determines how effectively that sealant will prevent moisture intrusion into the wall system and how the sealant will behave when under stresses such as compression, tension, or shear [1]. The adhesion of any sealant is highly dependent on the composition and conditions of the substrate.

Level of surface preparation--Levels of surface preparation can vary greatly in practice from what is specified. The level of joint preparation required can also greatly affect the cost of sealant installation, depending on the labor required. Sealant installations are also not always performed according to specifications. Primers are sometimes not used, the existing sealant may not be completely removed, or the joints may not be cleaned or ground to new material.

In this particular case, evaluation of the surface preparation required was critical because the original sealant adhesion was found to be greatly affected by the presence of moisture. This was evidenced by both the observed field conditions and by water immersion studies conducted in the laboratory on samples of original sealant. If the new sealant was applied over traces of the existing sealant which may be left after inadequate joint preparation, the bond of the new sealant could be adversely affected when the substrate material becomes saturated with water. Our testing had to be designed to account for as many of these conditions as possible to ensure a long-lasting application.

Performance in a wet environment--This project required the evaluation of the sealant in a wet environment due to the existing project conditions. The marble substrate was relatively thin and porous in comparison with other typical stone substrates, readily allowing moisture migration into and through the substrate adjacent to the new sealant. The sealant selected had to remain adhered and continue to perform effectively in this environment.

Compatibility with substrate--It is important to determine whether the existing substrate is compatible with the application of a new sealant. Occasionally a sealant may be adversely effected by the substrate, causing it to stain or deteriorate. Incompatibility can also potentially disrupt or inhibit bond of the new sealant, or prematurely deteriorate the sealant.

Compatibility with existing sealant--If a new sealant is to come into contact with the original sealant, it is important to know that the two sealants are compatible, so that the original sealant will not alter, degrade, or physically limit the performance of the new sealant. In this project, it was anticipated that some of the original sealant might not be completely removed from the joints to be resealed due to difficult field conditions. Therefore, it was important to confirm that the new sealant was compatible with the original sealant, to ensure that long-term adhesion or durability was not affected.

Durability and long-term performance--It is important to anticipate the long-term performance of a sealant, in order to evaluate its cost effectiveness relative to other sealant choices, and to assess the level of maintenance the sealant may require.

Substrate staining--Some sealants, or the primers required for installation, may stain or discolor the substrate. Porous, light-colored substrates are especially susceptible to this problem. Staining can adversely affect the aesthetics of project.

Constructability--Constructability issues affecting the installation of the sealant should be evaluated. These include mixing difficulty (impacts probability that proper mixing will occur), drying time of primer (may impact whether the joint primer can cure before installation of the sealant, thereby impacting productivity), tack-free and initial-set time (the longer the time to initial set, the more susceptible the sealant to attracting moisture and airborne dirt) and tooling characteristics (typically impacts durability and appearance).

## TESTING PROGRAM DESCRIPTION AND RESULTS

To address these critical issues affecting sealant performance, a sealant testing program was developed with three basic components: peel adhesion strength evaluation, accelerated weathering, and staining/plasticizer migration testing. These tests provided a basis for comparing the trial sealants to predict their potential performance if used on the subject project. The following is a summary of the phases of the testing program, the observations, and results of the tests.

### Identification of Existing Sealant

Work on the project began with the most critical steps in any testing program for a reseal application: identifying the type of sealant originally used in the wall system in order to assess the potential for incompatibility with a replacement sealant. In this case, the original sealant was identified by infra-red (IR) spectroscopy as an acrylic copolymer. The identification process requires solvent extractions to isolate the polymeric binder and plasticizing agent from the inorganic fillers and pigments present within the sealant matrix. In this case, samples of the original sealant were observed to be susceptible to volume swell and degradation after being immersed in water for seven days in the laboratory. The samples were visually examined and weighed before and after immersion.

### Selection of Trial Sealants

The next step was the selection of trial sealant products to be considered for the application. Due to budget and time constraints, the number of sealants that could be selected for testing was limited. Multi-component, non-sag urethane sealants were chosen for evaluation due to their long history of successful use with porous stone substrates. Final selection was based on our past experience with the products, documented performance on projects with a similar substrate, and the results of field adhesion and compatibility testing performed at the building site. Three urethane sealant products were selected. A silicone product was also added for comparison with the three polyurethane sealants, bringing the total number of trial sealants to four.

Obtaining Samples for Laboratory Testing

In order to accurately evaluate the joint sealant for a reseal application, the actual substrate must be used in the testing. Samples for testing were cut from representative marble panels removed from the wall system, crated and shipped to the laboratory for preparation. These samples represented the actual material that was to be the substrate for the replacement sealant.

Sample Preparation

The marble panels used for testing were cut into 48 prisms, approximately 7.6 cm (3 in.) wide by 15.2 cm (6 in.) long. The prisms were cut from the perimeter of the panels, to include a portion of the original sealant joint. A residual amount of the original sealant was left intact on one long side of each prism.

Parallel sets of samples with two levels of joint preparation were used to simulate a potential range of field conditions and test compatibility between the new and original sealant. Even if complete removal of the original sealant is specified, residue from the original sealant remains on the substrate surface in many instances, which may impact the performance of the new sealant. Substrate surface preparation is highly dependent on the installer and subject to deviations from what is specified. Therefore, twenty-four of the prisms were prepared by slicing away the original sealant, leaving a residue of the original on the surface of the joint. The remaining prisms were prepared by lapping (lapping is the process of removing all visible sealant and a 2 mm (1/16 in.) layer of the marble substrate to obtain a clean joint substrate free of any sealant or its absorbed components).

Prisms were paired in 24 wooden frames with set screws to hold the prisms firmly in place (Fig. 2). Unlapped prisms were paired together, as were the lapped prisms. Each pair of prisms was positioned to form a 13 mm (1/2 in.) joint for the new sealant. Segments of closed-cell backer rod were installed in the joints, against the backing of the frames, to maintain a consistent joint depth. Variations in depth were within 2 mm (1/16 in.) between pairs of prisms.

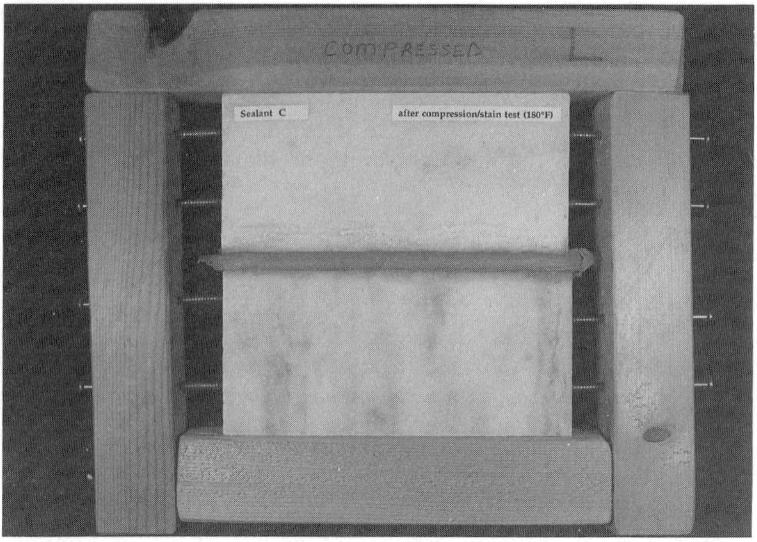


FIG. 2 -- Typical test set-up with two marble prisms in a wooden test frame. Sealant has been installed in the joint formed by the two prisms.

The joints between each pair of prisms were primed and sealed in accordance with manufacturers recommendations. The primers were chosen by each manufacturer for the marble substrate[4]. After the primers had dried, the sealant was installed, after mixing in accordance with manufacturer's directions, using a manual pump-action caulking gun. The joints were tooled to form concave beads, 10 mm (3/8 in.) deep at the centers. The four sealants were used in six pairs of prisms each. Three of the six pairs with each sealant were unlappped. The remaining three pairs were lapped. Table 1 summarizes the samples prepared for each sealant.

---

[4] The manufacturer of Sealant D initially did not recommend the use of a primer. After completion of the studies, the manufacturer of Sealant D claimed that a primer should have been used.

TABLE 1--SUMMARY OF SAMPLES FOR SEALANT TESTING PROGRAM

| Sealant Product | Sample No. | Joint Preparation | Peel-Adhesion Strength | Accelerated Weathering | Staining/ Plasticizer Migration |
|---|---|---|---|---|---|
| Sealant A | A1 | Unlapped | ■ | | |
| | A2 | Unlapped | | ■ | |
| | A3 | Unlapped | | | ■ |
| | A4 | Lapped | ■ | | |
| | A5 | Lapped | | ■ | |
| | A6 | Lapped | | | ■ |
| Sealant B | B1 | Unlapped | ■ | | |
| | B2 | Unlapped | | ■ | |
| | B3 | Unlapped | | | ■ |
| | B4 | Lapped | ■ | | |
| | B5 | Lapped | | ■ | |
| | B6 | Lapped | | | ■ |
| Sealant C | C1 | Unlapped | ■ | | |
| | C2 | Unlapped | | ■ | |
| | C3 | Unlapped | | | ■ |
| | C4 | Lapped | ■ | | |
| | C5 | Lapped | | ■ | |
| | C6 | Lapped | | | ■ |
| Sealant D | D1 | Unlapped | ■ | | |
| | D2 | Unlapped | | ■ | |
| | D3 | Unlapped | | | ■ |
| | D4 | Lapped | ■ | | |
| | D5 | Lapped | | ■ | |
| | D6 | Lapped | | | ■ |

The tack-free times ranged widely for the four sealants, as did the relative viscosity. The shortest tack-free time was 24 hours for Sealant A. All of the other sealants were found to be within 72 hours. This is a concern for the field installation since air and water-borne dirt could collect on the tacky joint surface and cause premature soiling of the sealant.

The initial set times for the four sealants also ranged greatly, from less than 24 hours to more than 48 hours for one of the sealants. The viscosity of the sealants also varied considerably, Sealant A being the most difficult to mix and Sealant B the easiest. Testing of the specimens was begun one week after installation.

## Mixing, Curing, Installation

During the mixing process, the relative viscosity of the four sealants was roughly gauged by difficulty of mixing. The sealants were allowed to cure for one week in the laboratory at ambient conditions of 72 °F and approximately 40% relative humidity. After 24 hours, the sealants were inspected to determine tackiness. Tack-free times and initial set were also noted. Tack-free times were roughly gauged by hand, with a spatula[5]. Table 2 outlines the mixing difficulty, as well as variations in initial set and tack-free times for the four sealants.

**TABLE 2--OBSERVATIONS DURING SEALANT MIXING, CURING, AND INSTALLATION**

| Sealant Product | Initial Set Time | Tack-Free Time[6] | Viscosity (Difficulty of Mixing) |
|---|---|---|---|
| Sealant A | 24 hours | 24 hours | high |
| Sealant B | 24 hours | 48 hours | low |
| Sealant C | 48 hours | 72 hours | medium |
| Sealant D | 24 hours | 48 hours | medium |

[5] The test procedure, ASTM C679, "Standard Test Method for Tack-Free Time of Elastomeric Sealants," was not used for these studies due to budget constraints.

[6] Some of the sealants remained tacky throughout the testing. The time noted therefore represents the point at which tackiness became consistent for the sample.

Initial Peel Adhesion Strengths (Wet Conditions)

Adhesion is typically field-tested in the industry by cutting a section of sealant free from the substrate and using it as a tab to attempt to pull the remaining bonded sealant from the substrate [2]. The objective of the test is to determine the relative amount of force required to cause sealant failure (indicating bond strength and adhesion), and the type of failure (either cohesive or adhesive). A cohesive failure indicates that the sealant adhesion to the substrate is adequate to develop the tensile capacity of the sealant. Elongation of the sealant as it is pulled (also measured) indicates the sealant's ability to accommodate movement of the joint substrate. The testing should be conducted under the worst conditions anticipated for the sealant during its service life. In this testing program, it was determined that the worst case conditions for the sealant would occur when the joint substrate was saturated with rainfall. The initial adhesion testing was therefore conducted after immersing the specimens in water.

Eight of the 24 composite marble/sealant specimens consisting of one lapped and one unlapped specimen of each sealant were prepared for initial adhesion testing by removing them from the wooden frames and immersing them in room-temperature tap water for one week. After soaking, the sealant beads were cut through the center of the joint and tested for peel adhesion strength by pulling, with a force gauge at a right angle to the marble surface. All the pull test specimens, except specimens for sealants with the lowest pull test values, elongated a minimum of 500% before either adhesive or cohesive failure. Table 3 summarizes the test data obtained for each of the specimens.

TABLE 3--RESULTS OF PEEL-ADHESION STRENGTH TESTS AFTER WATER IMMERSION

| Sealant Product | Joint Surface Preparation | Sample No. | Adhesion Force (N/Lbs) | Mode of Failure[7] |
|---|---|---|---|---|
| Sealant A | Unlapped | A1 | 35.6/8 | Cohesive |
| | Unlapped | | 62.3/14 | Cohesive |
| | Lapped | A4 | 35.6/8 | Cohesive |
| | Lapped | | 48.9/11 | Cohesive |
| Sealant B | Unlapped | B1 | 4.5/1 | Adhesive |
| | Unlapped | | 40.0/9 | Adhesive |
| | Lapped | B4 | 31.1/7 | Adhesive |
| | Lapped | | 53.4/12 | Adhesive |
| Sealant C | Unlapped | C1 | 17.8/4 | Partial Adhesive |
| | Unlapped | | 17.8/4 | Partial Adhesive |
| | Lapped | C4 | 17.8/4 | Cohesive |
| | Lapped | | 17.8/4 | Cohesive |
| Sealant D | Unlapped | D1 | 17.8/4 | Adhesive |
| | Unlapped | | 17.8/4 | Adhesive |
| | Lapped | D4 | 8.9/2 | Adhesive |
| | Lapped | | 13.3/3 | Adhesive |

Based on the data obtained, Sealant A performed better than the other three sealants, according to the type of failure and the level of force required to fail the sealant. Sealant B, C, and D experienced mostly adhesive failures, indicating insufficient bond strength to develop the full tensile capacity of the sealant.

Sealant A performed without any adhesive failures in the laboratory adhesion testing. The adhesion forces were the highest values obtained from any of the sealants. Adhesion force values were similar for the lapped and unlapped joint surface samples. This indicates that adhesion of Sealant A would be adequate under the varying levels of joint preparation anticipated in the field.

---

[7] Adhesive indicates that failure occurred at the interface between the sealant and the marble substrate. Cohesive indicates that failure occurred entirely within the sealant bead, with no adhesive failure from the substrate. Partial Adhesive indicates that some partial adhesive failure occurred prior to a cohesive failure.

During the laboratory adhesion testing, Sealant C experienced partial adhesive failures on the marble sample with the unlapped joint surface. Cohesive failures were obtained with the lapped joint surface, indicating that the sealant performed better with a higher degree of joint preparation. The adhesive-mode failures indicated that some incompatibility may exist between Sealant C and the existing sealant, or that water immersion may have affected the bonding of the sealant. The adhesion force required to fail Sealant C was also low, indicating low bond strength with the marble substrate in comparison to Sealant A and Sealant B.

Staining/Plasticizer Migration

Staining of a porous substrate is commonly caused by migration of a liquid component, or plasticizer, in the sealant. The migration of plasticizer has been found to be more likely to occur under high temperatures, as experienced by exterior building joints in summer months. Migration is also more likely when the joints are compressed due to movement or expansion of the substrate, forcing the plasticizer out of the sealant and into the substrate.[8] The test for staining/plasticizer migration simulated these conditions using the four sealants and the actual substrate material.

Eight of the 24 composite marble/sealant specimens, consisting of one lapped and one unlapped specimen of each sealant were selected for compression within stainless steel frames. After removal of the backer rod, the sealant joints were compressed 50%, to a width of 6 mm (1/4 in.). The compressed samples were heated to 160 °F for two weeks, and then to 180 °F for another two weeks. Table 4 summarizes the results and observations of the samples.

---

[8] ASTM Test Method for Staining of Porous Substrates by Joint Sealants (C 1248), was under development at the time of these studies. The C 1248 procedure involves compression of all specimens, and exposure of some specimens to temperatures of 158 °F for up to 28 days.

TABLE 4--SUMMARY OF OBSERVATIONS FOR STAINING/PLASTICIZER MIGRATION TESTING

| Sealant Product | Sample No. | Joint Surface Preparation | Staining/Plasticizer Migration Testing | |
|---|---|---|---|---|
| | | | 2 Weeks at 160°F | 2 Weeks at 180°F |
| Sealant A | A3 | Unlapped | Very slight darkening of marble adjacent to sealant, approx. 3 mm (1/8 in.) each side of joint. | No changes. |
| | A6 | Lapped | | |
| Sealant B | B3 | Unlapped | Slight darkening of marble adjacent to sealant, approx. 3 mm (1/8 in.) each side of joint. | Darkening was most distinct of all samples. Formerly elastomeric sealant became tacky, and fluid, with complete loss of adhesion. |
| | B6 | Lapped | | |
| Sealant C | C3 | Unlapped | Slight darkening of marble adjacent to sealant, approx. 2 mm (1/16 in.) each side of joint. | Darkening of marble increased to approx. 3 mm (1/8 in.) each side of joint. |
| | C6 | Lapped | | |
| Sealant D | D3 | Unlapped | Sealant changed color from grey to light brown. | No changes. |
| | D6 | Lapped | | |

After one week under compression, the samples did not exhibit any plasticizer migration from sealant to marble. After two weeks at 160 °F, a slight darkening of the white marble substrate adjacent to the joint was detected in samples from three of the four sealants. Also, the surface of Sealant D had changed color from light grey to light brown[9]. Similar observations were made after the samples were exposed to 180 °F for another two weeks, except that the staining present in the Sealant B samples was more distinct, and the staining of Sealant C appeared to have widened (Figs. 3 and 4). After cutting through the samples that exhibited the staining to examine the depth of the stain, it was determined that the staining was due to plasticizer migration. When the samples were cut, it was observed that Sealant B had experienced a complete loss of cohesion during the test. The sealant material reverted to a fluid, tacky state (Fig. 5).

---

[9] The manufacturer of Sealant D stated after the testing that the color change would not occur in an exterior application.

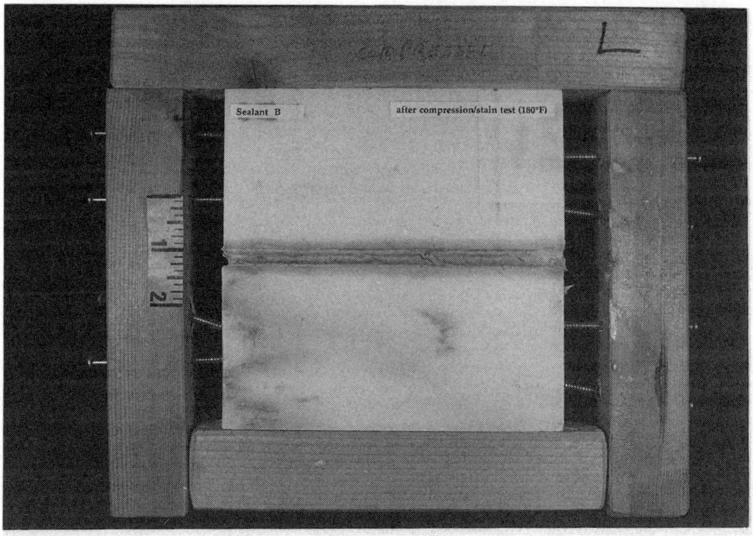


FIG. 3 -- Sealant B Sample frame. Note the uniform staining adjacent to the sealant.

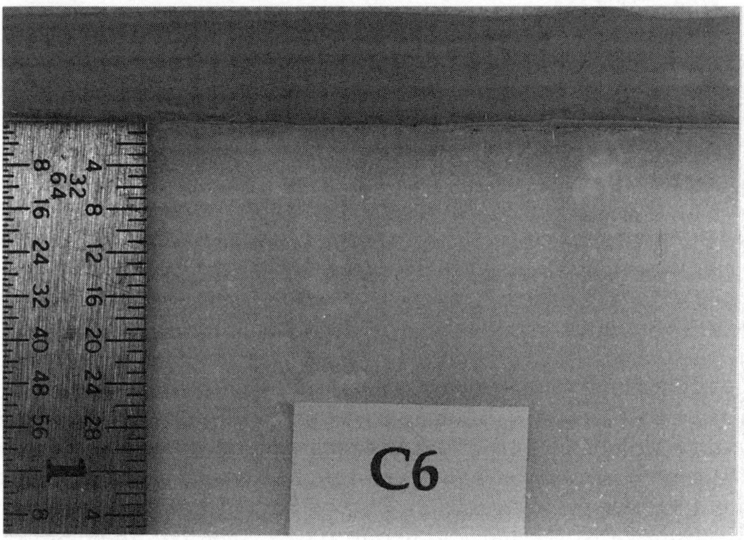


FIG. 4 -- Sealant Sample C. Note the staining adjacent to the joint sealant.

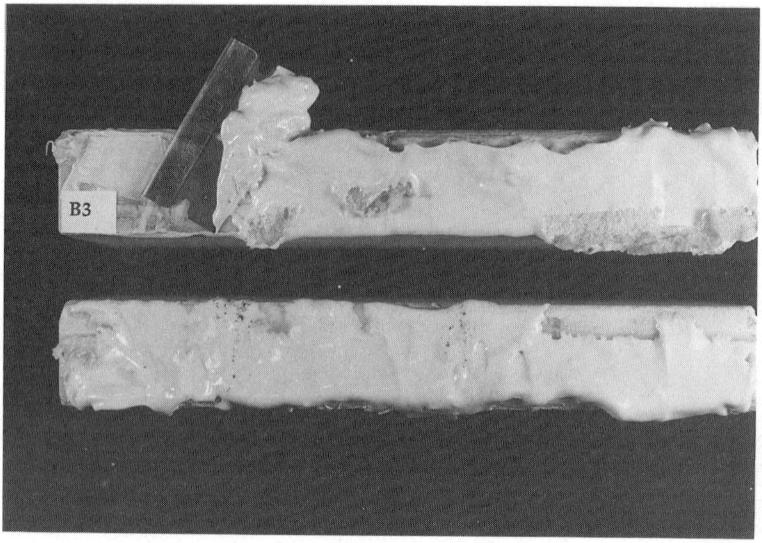


FIG. 5 -- Sealant B Sample after Staining/Plasticizer Migration Testing. Sealant sample has reverted to a fluid and tacky state.

Accelerated Weathering

Accelerated weathering provides data with which to evaluate the potential long term performance of a sealant. Accelerated weathering allows the sealant to be exposed to conditions similar to those anticipated after actual installation, but in a short period of time. This test therefore can give an indication of sealant performance over time. Accelerated weathering should only be considered as a method of comparing sealant durability, not an absolute time period of outdoor exposure. Only very rough correlations between weatherometer hours and actual outdoor exposure are possible due to the high number of variables, such as site specific climate data, geographic location of the project site, and variations in weatherometer equipment.

The weatherometer used for the testing is an Atlas, enclosed arc weatherometer. The apparatus exposes the specimens to intense ultraviolet (UV) light with intermittent water spray. The UV light sources in our apparatus are twin enclosed carbon arc lamps. The apparatus conforms to the ASTM Practice for Operating Light-Exposure Apparatus, Carbon Arc Type, with or without Water for Exposure of Non-metallic Materials (G 23, Type D). In-house correlations indicate that 1000 hours in the Atlas weatherometer is roughly equivalent to one or two years of outdoor exposure in the Washington, D.C. area.

The test was designed to follow the exposure conditions given in ASTM Standard Test Method for Effects of Accelerated Weathering on Elastomeric Joint Sealants (C 793), except that the sample configuration was slightly different, and the exposure time lengthened from the specified 250 hours to 1000 hours. The length of exposure was limited by project budget. Samples were viewed at various intervals to determine points at which the sealant was affected by the test conditions.

For exposure to the weatherometer, eight of the 24 composite marble/sealant specimens, consisting of one lapped and one unlapped specimen of each sealant, were removed from the wooden frames and fastened into stainless steel frames. Table 5 summarizes the observations made at 100 hours, 260 hours, and 1000 hours of exposure.

**TABLE 5--SUMMARY OF OBSERVATIONS FOR ACCELERATED WEATHERING**

| Sealant Product | Sample No. | Joint Surface Preparation | Accelerated Weathering | | |
|---|---|---|---|---|---|
| | | | 100 Hours | 260 Hours | 1000 Hours |
| Sealant A | A2<br>A5 | Unlapped<br>Lapped | No deterioration. | Brown stains on marble adjacent joint, 3 mm (1/8 in.) wide each side of joint, possibly due to primer discoloration. | Same as 260 hour observations. |
| Sealant B | B2<br>B5 | Unlapped<br>Lapped | No deterioration. | No deterioration. | No deterioration. |
| Sealant C | C2<br>C5 | Unlapped<br>Lapped | Brown stains on marble adjacent to joint, appeared to be discoloration of primer. | Increased brown stains on marble adjacent to joint 3 mm (1/8 in.) wide each side of joint, possibly due to primer discoloration. | Same as 260 hour observations. Some surface cracking of sealant, up to 2 mm (1/16 in.) deep. |
| Sealant D | D2<br>D5 | Unlapped<br>Lapped | No deterioration. | No deterioration. | No deterioration. |

The sealants tested appeared to withstand the weathering without degradation except for some cracking on the surface of the Sealant C samples (Fig. 6). The primers for Sealant A and C became discolored, causing irregular dark-brown stains on the white marble surface due to exposure to UV light (Fig. 7). The stains penetrated into the stone less than 1 mil. Preliminary indications are that when the primer is exposed to environmental conditions in the field, it may tend to discolor over time. Control of the primer application is recommended by masking or other means to avoid excess primer on the surface of the stone and minimize the discoloration is recommended, however this may be difficult to control in an actual installation.

FIG.6 -- Sealant Sample C. Surface cracking was evident after 250 hours of weatherometer exposure.

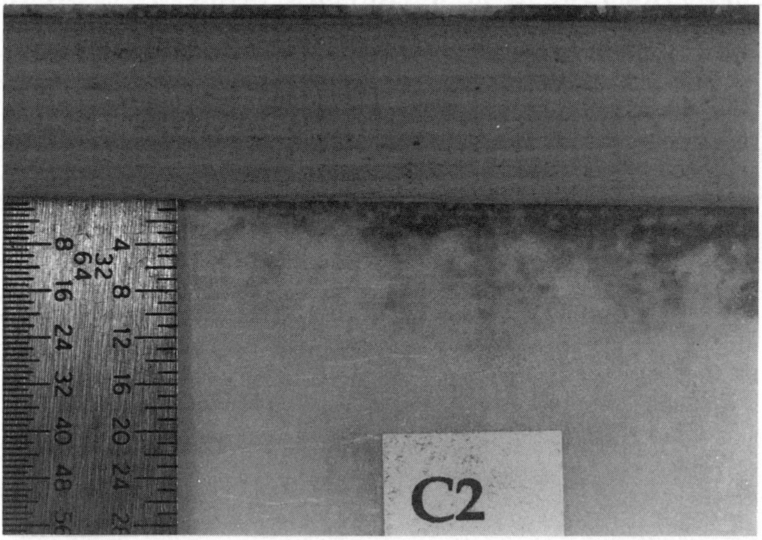


FIG. 7 -- Sealant C Sample. Dark brown stains were observed on the marble surface, due to primer discoloration.

Peel-Adhesion Strength after Accelerated Weathering

To obtain more adhesion data, the specimens previously subjected to accelerated weathering were also tested for adhesion. Not only did this provide more data on relative adhesion properties with the two types of surface preparation, it also allowed for study of adhesion characteristics after extensive simulated weathering of the new sealant.

Adhesion tests were performed in the same manner as the initial peel-adhesion strength tests, except that the samples were not soaked before testing. The force required to fail the sealant and the type of failure were noted. Table 6 summarizes the results.

**TABLE 6--RESULTS OF PEEL-ADHESION STRENGTH TESTS OF ACCELERATED WEATHERING SAMPLES**

| Sealant Product | Joint Surface Preparation | Sample No. | Adhesion Force (N/Lbs) | Mode of Failure |
|---|---|---|---|---|
| Sealant A | Unlapped | A2 | 22.2/5 | Cohesive |
| | Unlapped | | 40.0/9 | Cohesive |
| | Lapped | A5 | 35.6/8 | Cohesive |
| | Lapped | | 35.6/8 | Cohesive |
| Sealant B | Unlapped | B2 | 31.1/7 | Partial Adhesive |
| | Unlapped | | 48.9/11 | Partial Adhesive |
| | Lapped | B5 | 26.7/6 | Partial Adhesive |
| | Lapped | | 44.5/10 | Partial Adhesive |
| Sealant C | Unlapped | C2 | 35.6/8 | Partial Adhesive[10] |
| | Unlapped | | 40.0/9 | Partial Adhesive[10] |
| | Lapped | C5 | 35.6/8 | Partial Adhesive[10] |
| | Lapped | | 35.6/8 | Partial Adhesive[10] |
| Sealant D | Unlapped | D2 | 26.7/6 | Cohesive |
| | Unlapped | | 22.2/5 | Cohesive |
| | Lapped | D5 | 22.2/5 | Cohesive |
| | Lapped | | 26.7/6 | Cohesive |

[10] Partial adhesive failure occurred at the outer edge of sealant/marble interface, adjacent to weathered surface, to a depth of approximately 1/8 in.

Peel-Adhesion Strength after Staining/Plasticizer Migration Testing

To obtain more adhesion data, the specimens previously subjected to joint compression as part of the staining/plasticizer migration tests were also tested for adhesion. This also provided more data relative to adhesion properties with the two types of surface preparation, and allowed for study of adhesion characteristics after the joints had undergone extensive compression. Adhesion tests were performed in the same manner as the initial peel-adhesion strength tests, but without soaking, yielded the results summarized in Table 7.

**TABLE 7--RESULTS OF PEEL-ADHESION STRENGTH TESTS OF STAINING/ PLASTICIZER MIGRATION SAMPLES**

| Sealant Product | Joint Surface Preparation | Sample No. | Adhesion Force (N/Lbs) | Mode of Failure |
|---|---|---|---|---|
| Sealant A | Unlapped | A3 | 35.6/8 | Cohesive |
| | Unlapped | | 26.7/6 | Cohesive |
| | Lapped | A6 | 31.1/7 | Cohesive |
| | Lapped | | 26.7/6 | Cohesive |
| Sealant B | Unlapped | B3 | * | Cohesive |
| | Unlapped | | * | Cohesive |
| | Lapped | B6 | * | Cohesive |
| | Lapped | | * | Cohesive |
| Sealant C | Unlapped | C3 | 44.5/10 | Cohesive |
| | Unlapped | | 31.1/7 | Cohesive |
| | Lapped | C6 | 44.5/10 | Cohesive |
| | Lapped | | 31.1/7 | Cohesive |
| Sealant D | Unlapped | D3 | 31.1/7 | Cohesive |
| | Unlapped | | 26.7/6 | Cohesive |
| | Lapped | D6 | 44.5/10 | Partial Adhesive |
| | Lapped | | 35.6/8 | Cohesive |

* Not measured due to reversion of samples to a tacky, fluid state after Staining/Plasticizer Migration Testing.

## CONCLUSIONS

This sealant testing program has highlighted some specialized methods that can be used for testing sealants with a marble substrate. It has demonstrated that important information can be obtained about the performance of sealants with a particular substrate in advance of actual installation, which can be essential to the designer and specifier when choosing a sealant product.

The sealant testing program described above also identified several key factors and concerns regarding the application of the trial sealants to the marble substrate. Some of these are as follows:

1. All of the sealants tested were compatible with the marble substrate. Throughout the testing, there was no evidence of adverse reactions between the stone and sealant, even after accelerated weathering.

2. Adhesion of the sealant to the substrate varied among sealants. The differences in adhesion performance could be demonstrated by both the mode of failure (adhesive or cohesive) and quantitatively by the force required to fail the sealant. The largest sampling possible is recommended to draw definitive conclusions regarding the relative adhesion performance of one sealant over another.

3. The residual acrylic sealant on the unlapped samples did not appear to have a significant impact on sealant adhesion as expected. The effect of surface preparation on the levels of adhesion appeared to be dependent on the individual sealant. The sealant that performed well in the adhesion tests were not appreciably affected by the variation in surface preparation.

4. Good initial adhesion test results were obtained for some of the samples after water immersion and saturation of the substrate. This indicated that the sealants would remain adhered, even under severe weather conditions.

5. From comparison of the quantity of adhesive-mode failures and the adhesion force values obtained during the three series of adhesion tests (water immersion, accelerated weathering, and joint compression), it appeared that the most rigorous test condition for sealant adhesion was the 7-day water immersion.

6. Accelerated weathering appeared to impact adhesion performance in Sealant C, by causing adhesive failure of the sealant from the substrate, near the exposed surface to a 3 mm (1/8 in.) depth. The sealant remained adhered below the surface.

7. The accelerated weathering indicated very little difference in sealant surface deterioration among the sealants tested, except for Sealant C which experiencing some cracking on the sealant surface. The increased deterioration is an indication that the sealant may have a shorter service life than the other sealants tested.

8. The accelerated weathering indicated the potential for primer discoloration with Sealants A and C. By identifying this problem through testing, the problem can be corrected before becoming a major concern during or after construction. The staining could be eliminated before construction by changing the installation procedures to limit the amount of primer on the substrate surface, or working with the manufacturer to select a different primer.

9. Results of the plasticizer migration studies showed that all the sealants, except Sealant D, had the potential to stain the substrate with plasticizer.

10. It was found that Sealant B became fluid, tacky, and lost all strength after exposure to 180 °F during the staining/plasticizer migration studies. This indicates the potential for instability of this sealant in environments exposed to extreme heat and substrate temperatures.

11. Sample preparation, revealed that Sealant C had a significantly longer initial setting time and tack-free time. Sealants B and C never became completely tack-free. A longer initial-set time increases the chances of the joint being disrupted, damaged, or prematurely soiled as dirt is retained for a longer period of time on the tacky sealant surface.

Although the testing program described was designed to address the concerns with a marble substrate, similar testing programs could be employed for a variety of substrates, with specialized testing developed to evaluate unique conditions of any sealant installation. Laboratory testing should be supplemented with field testing and mock-ups whenever possible, as well as available manufacturer's test data for similar substrates.

## REFERENCES

[1] American Concrete Institute (ACI), ACI Committee 504, "Guide to Sealing Joints in Concrete Structures," ACI Manual of Construction Practice, American Concrete Institute, Detroit, MI, 1992.

[2] Sealant, Waterproofing & Restoration Institute (SWRI), "SEALANTS: THE Professionals' Guide," Sealant, Waterproofing & Restoration Institute, Kansas City, MO, 1990.

# Advances in Adhesion Testing and Adhesive Promoters

Anthony P. Cerra[1] and Voytek S. Gutowski[2]

# PERFORMANCE-BASED ADHESION TESTING OF STRUCTURAL SEALANTS

**REFERENCE:** Cerra, A.P., and Gutowski, W.S., **"Performance-based Adhesion Testing on Structural Sealants,"** Science and Technology of Building Seals, Sealants, Glazing, and Waterproofing: Fifth Volume, ASTM STP 1271, Michael A. Lacasse, Ed., American Society for Testing and Materials, 1996.

**ABSTRACT:** Existing accelerated laboratory procedures for the assessment of the adhesion of silicone sealants fail to consider the fact that the sealant/substrate interface in a building facade is continuously subjected to a combination of both mechanical and environmental stress. As a result of this, standardised test protocols lack the discrimination necessary to assess relative adhesive behaviour. A novel test procedure is offered which enables improved discrimination of the adhesive strength of the sealant/substrate system. The method is based on the simultaneous application of both mechanical and hydrothermal stress, in order to promote adhesive failure at loads below the cohesive strength of the bulk sealant.

A simpler version of the novel procedure allows for the screening of sealant/substrate combinations that represent potential long-term durability risks. This is achieved by the combination of hydrothermal and mechanical stress at a level close to the structural silicone design stress (0.138 MPa) applied through imposing a 12.5% strain (equal to the normal design strain in silicone sealants) together with immersion of the joint in hot water.

**KEYWORDS:** sealants, silicone, adhesion, testing, accelerated, structural glazing.

## INTRODUCTION

The main function of structural silicone adhesives (sealants) is to permanently bond external cladding panels to the facade frame, and to ensure the structural integrity and safety of the building envelope throughout its entire service life. Equally important is the provision of weathertightness for the building.

[1]Experimental Scientist and [2]Senior Principal Research Scientist, respectively, CSIRO Division of Building, Construction and Engineering, PO Box 56, Highett, Victoria 3190, Australia.

The chemical structure of silicone sealants ensures chemical bonding across the sealant/substrate interface at the time of the facade fabrication. The question remains, however, whether the quality and integrity of this bonding manifested by the strength of the sealant/substrate adhesion, will be retained during the expected life of the building.

The reliance of the facade integrity and structural safety on adhesive bonding alone, and the relatively young age of this technology, require a thorough understanding of the bond behaviour. In particular, the influence of the service environment on the possible bond deterioration requires careful consideration and realistic simulation in order to develop predictive models and accelerated laboratory procedures for the assessment of the expected long-term performance.

There are numerous US and European standards for the evaluation of elastomeric sealants generally and, in some cases, they apply to silicone sealants specifically. However, most of these methods were originally developed to assess the behaviour of these materials when applied as sealants only and not as load-bearing structural adhesives. The general strategy adopted in these standards has been to condition the specimens in a predetermined sequence of various environmental conditions prior to actual mechanical testing. The results such as bond strength and percentage adhesive failure (% AF) (delamination of the sealant/substrate interface) constitute the basis for the assessment of the sealant/substrate system for a particular application. For example, the major structural silicone glazing (SSG) sealant suppliers in the USA have specified an ASTM C794 Peel Test [1] on specimens that have been conditioned for one week in water at room temperature. However, it is generally accepted that the peel test is only suitable for the qualitative assessment of failure mode and that a tensile test is necessary for the quantitative assessment of bond strength [2, p.97]. Alternatively the Europeans (e.g. CSTB in France) have preferred a tensile test after a pre-conditioning regime involving both hot air exposure at 100°C and water at room temperature [3]. It is significant that ASTM has very recently published specification C1135 "Test Method for Determining Tensile-Adhesion Properties of Structural Silicone Sealants". One important difference between these two procedures is that the French specification allows for two days drying time after water immersion, whereas ASTM specifies a maximum of only 10 minutes delay between the removal of the specimen from water and testing. Considering the dynamic nature of the silicone/substrate bond in the presence of water [2, p.295], this discrepancy could prove to be very significant.

Furthermore, current methods [3] yield only semi-quantitative data. For example, the result for a shear, peel or tensile test after preconditioning will consist of a % AF as well as an ultimate strength value. Estimating the former can be subjective as the actual nature of the failure (cohesive or adhesive) is often open to interpretation, particularly in those cases involving cohesive failure close to the bondline [4]. The ultimate strength value has several disadvantages as a measure of the adhesive performance of a sealant, except in cases where unequivocal 100% AF occurs. In instances of partial adhesive failure, the peak load is often a measure of both the energy required to create and propagate the adhesive failure and the cohesive strength of the bulk sealant. The reason for this is that in conventional testing the applied stress is dynamic and increases continually until complete failure of the joint via crack propagation at the interface or through the bulk sealant. Therefore in cases of combined failure modes, the reduction in strength due to adhesive breakdown may be masked. It would be more meaningful to test at constant stress levels, well below the cohesive strength of the sealant, so as to ensure that failure is initiated at the bondline.

It has been recognised by the first author of this paper [5] that existing procedures for the assessment of adhesive performance of silicone sealants fail to consider the following:

(a) the sealant/substrate interface in a building facade is continuously subjected to a complex stress field arising from the tensile and shear deformation of the sealant due to the negative wind pressure and thermal movements of cladding versus framing;

(b) the most detrimental environmental factor leading to the deterioration of any adhesive bond is moisture which is present at the sealant/substrate interface in the form of vapour or a condensed film; and

(c) the chemical bonds between the silanol groups of the silicone sealant and the surface chemical groups of the substrate (e.g. hydroxyl-, carboxyl-, carbonyl-) may be either a hybrid ionic-covalent oxane bonds or hydrogen bonds [4, p.598]. These interfacial bonds may be constantly broken and re-formed by reversible hydrolysis.

Therefore one may assume that at an unloaded interface in the presence of moisture, the reactive groups of the sealant and substrate are held in such proximity that free silanol groups created as a result of the hydrolysis of interfacial bonds can be frequently re-formed to the original bonds or can make new bonds with the available adjacent surface groups on removal of moisture from the interface. If, however, the presence of sufficient strain in the flexible silicone resin results in the hydrolised interfacial bonds being moved away from the substrate surface, no re-forming of the bonds will occur. Subsequent to this, progressive adhesive failure along the sealant/substrate interface will propagate under the combined influence of surface moisture and the tensile or shear strain imposed on the sealant bead.

Thus, the basic strategy underlying the development of the novel testing procedure described in this paper involved the application of both mechanical and environmental (hydrothermal) stress simultaneously rather than sequentially, thereby realistically simulating in-service conditions and inducing adhesive failures at stress levels below the cohesive strength of the sealant. The mechanical stress (tensile for the sake of simplicity) is applied at a realistic level in the context of current design practice for SSG, i.e below or slightly above 0.138 MPa (20 psi).

Therefore, an improved adhesion performance test has been developed both for comparing existing materials and for the development of new products. The basic philosophy behind this performance test, and ultimately a performance specification, is that the test evaluates the property being assessed, so that a minimum performance level may be specified. The test is specifically designed to provide an accelerated procedure directly examining adhesion at the substrate/sealant interface.

## EXPERIMENTAL DETAILS

### Sealants and Substrate Materials

Three one-component RTV silicone sealants from three manufacturers were used in experiments. Sealants denoted 1 and 38 were alkoxy-cured and sealant 18 was an oxime-cured material.

A description of all the sealant/substrate combinations tested is provided in Table 1. These materials were selected as being representative of ones typically prescribed for modern building facades designed with curtain walls, and employing the use of SSG in its various forms.

Tensile specimens with 40 × 12 × 12 mm sealant beads were prepared and allowed to cure for 2 months at standard laboratory conditions (i.e., 25°C and 50% RH) prior to testing. All substrates were cleaned with methyl ethyl ketone prior to sealant application.

TABLE 1—Combinations of materials tested.

| Combination no. | Substrate code | Substrate description | Sealant no.[a] |
|---|---|---|---|
| 1 | B11 | Black anodised accelerated seal aluminium (4 mm) | 1 |
| 2 | PBr | Brown 50% gloss polyester powder-coated aluminium | 1 |
| 3 | M9 | Green granite (19 mm) | 1 |
| 4 | M9 | as above | 1 primed |
| 5 | P5 | Matt black polyester powder-coated aluminium | 1 |
| 6 | M6 | Compressed fibre cement sheet (19 mm) | 1 primed |
| 7 | P1 | Gloss white polyester powder-coated aluminium | 18 |
| 8 | B6 | Black anodised cold-sealed aluminium | 18 |
| 9 | G10 | Stainless steel-coated reflective glass (6 mm) | 18 |
| 10 | D1 | White fluorocarbon-coated aluminium | 18 |
| 11 | G10 | as above | 38 |
| 12 | G1 | Annealed clear float glass (6 mm) | 38 |
| 13 | D1 | as above | 1 |
| 14 | DB5 | Dark bronze anodised, hot-sealed aluminium | 1 |
| 15 | M2 | Developmental modified fluorocarbon-coated aluminium | 1 |
| 16 | B9 | Black anodised, accelerated-seal aluminium | 1 |
| 17 | EB3 | Black anodised aluminium ("Chemel" process) | 1 |
| 18 | EA1 | Clear anodised aluminium ("Chemel" process) | 1 |
| 19 | EB3 | as above | 18 |
| 20 | EA1 | as above | 18 |
| 21 | DB5 | as above | 18 |
| 22 | P12 | High durability grey polyester powder-coated aluminium | 18 |
| 23 | M2 | as above | 18 |
| 24 | B9 | as above | 18 |

[a] All sealants are commercial architectural grade one-part, black, neutral-cure, RTV silicones from a range of manufacturers. All were used unprimed except for combinations 4 and 6 as shown. The recommended commercial primers were used to prime the substrates where indicated.

### Stress-Strain Tests During Immersion

Mechanical tests in hot tap water were conducted using a special bath attached to a mechanical testing machine. This allowed stress-strain testing during immersion with the water temperature being thermostatically controlled to ±1°C.

Tensile specimens (40 × 12 × 12 mm) were tested in triplicate at standard laboratory conditions and at water temperatures of 23, 40, 55 and 70°C. All specimens were conditioned for 1 hour in the water bath, at the particular test temperature, prior to testing at a crosshead speed of 5 mm/minute.

### Constant Strain Tests During Immersion

Another part of the investigation was to apply constant strain during immersion in hot water. Tensile specimens with 40 × 12 × 12 mm sealant beads were used and these were "blocked" with polyethylene and metal spacers to achieve the required strain, as shown in Fig. 1. Strains of 12.5 and 25% were used in combination with water temperatures of 40, 55 and 70°C and the % AF was visually assessed during the test. Tensile tests at 50 mm/min were conducted on those specimens that did not fail completely during the exposure period. Specimens were not allowed to dry out prior to tensile testing. Combinations 1 to 12 only, as specified in Table 1, were tested under conditions of constant strain and all tests were done in triplicate.

Constant Stress Tests During Immersion

These experiments were conducted by applying dead loads to tensile specimens with 40 × 12 × 12 mm sealant beads while they were immersed in a heated water tank. This was achieved by designing a frame around the tank so that weights could be hung outside it by means of tension lines (see Fig. 2). All the combinations described in Table 1 were tested. The load required to fail the joint within 24 hours immersion forms the basis of the result, and this load is varied according to a special statistical procedure which is described in detail by Dixon [6].

## RESULTS AND DISCUSSION

Stress-Strain Tests During Immersion

Figures 3–5 illustrate the stress-strain characteristics of sealants 1, 18 and 38 respectively. Three important observations can be made from these results:

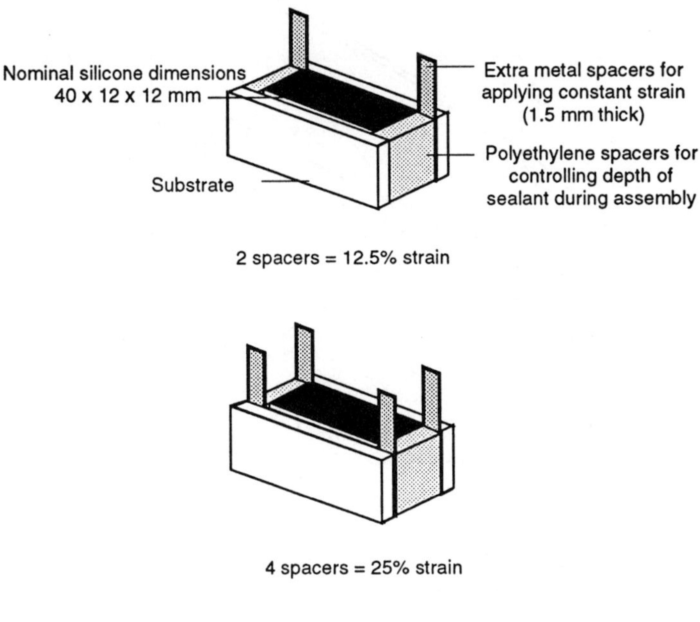


FIG. 1—Design of constant strain specimens.

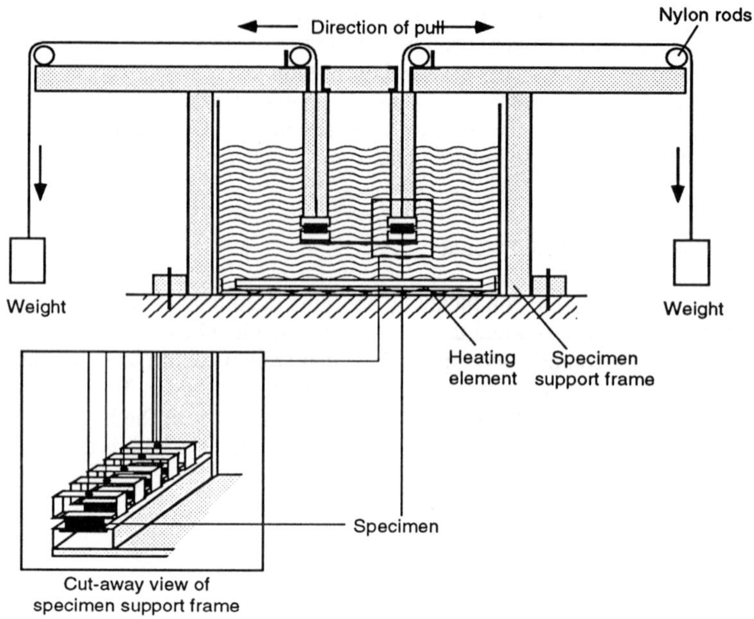


FIG. 2—Design of apparatus for applying dead loads to tensile specimens during hot water immersion.

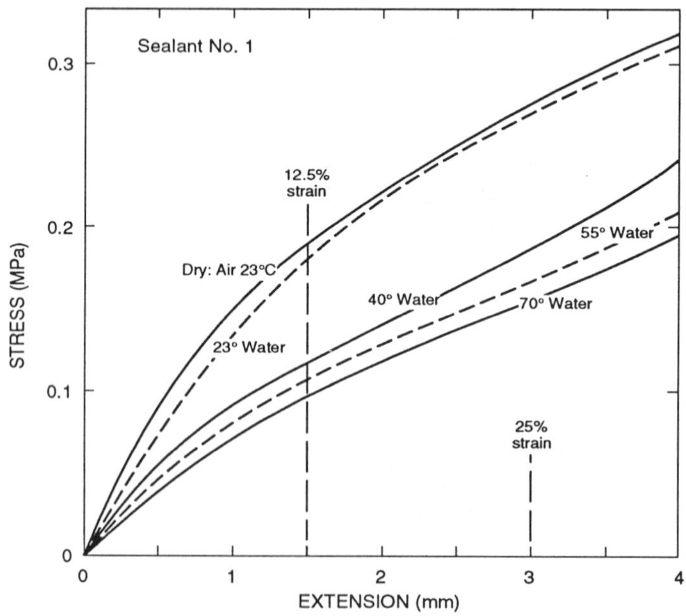


FIG. 3—Stress-strain characteristics of sealant 1 tested in dry air at 23°C, and in water at 23, 40, 55 and 70°C.

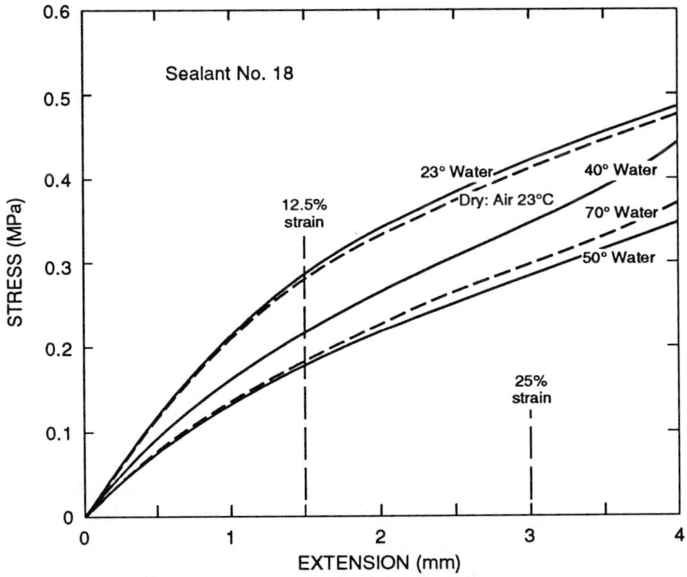


FIG. 4—Stress-strain characteristics of sealant 18 tested in dry air at 23°C, and in water at 23, 40, 55 and 70°C.

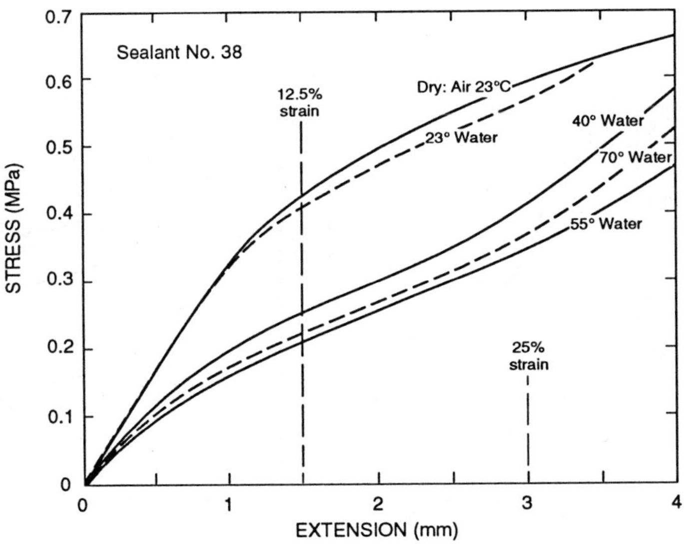


FIG. 5—Stress-strain characteristics of sealant 38 tested in dry air at 23°C, and in water at 23, 40, 55 and 70°C.

(a) the stress-strain curves of all sealants conditioned for 1 hour in 23°C water before testing while immersed, are very similar to the results obtained in tests conducted in air at 23°C;

(b) an increase in water temperature from 23°C to 70°C results in a rapid plasticisation of all the sealants tested, as shown by the reduction in stress at given strain when compared with the dry specimens or those conditioned in 23°C water; and

(c) as seen in Fig. 6, an increase of water temperature results in a semiparabolic decrease of the stress level for a given strain, (second-order relationship "stress versus water temperature" within the scope of the current experiment), with a stress "plateau" approached at approximately 55°C for 12.5% strain and about 70°C for 25% strain.

In Fig. 6 we have incorporated a reference line indicating the nominal value of the commonly accepted tensile design stress for structural sealants, i.e 0.138 MPa (20 psi). Also incorporated in this figure is an axis (right-hand side) translating the stress level into the actual value of the design safety factor (multiples of 0.138 MPa).

It is clearly seen from Fig. 6, that the average stress in sealants subjected to a simultaneous tensile strain of 12.5% (which is a typical design strain for structural sealants) and immersion for instance in 55°C water approaches a level of about 0.7, 1.3 and 1.6 of the design stress for sealants 1, 18 and 38 respectively. Based on this observation, it can be implied that any sealant/substrate combination that would fail by interfacial delamination when subjected to immersion in hot water at 12.5% constant strain may not be acceptable for structural glazing applications, due to the possibility of failure occurring at an unacceptably low safety factor.

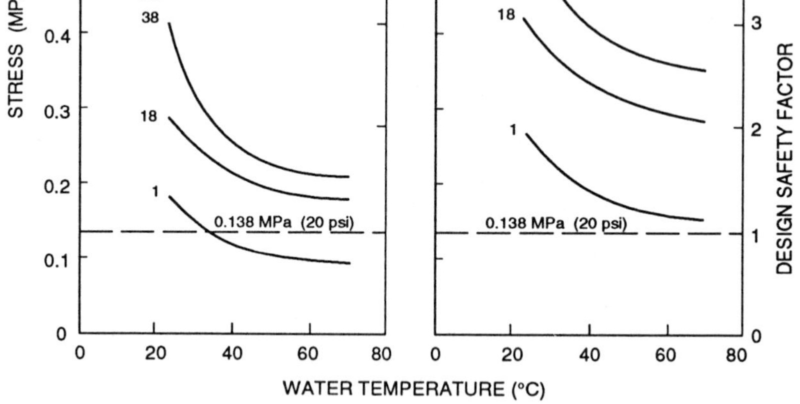


FIG. 6—Average stress level in sealants 1, 18 and 38 at two strain levels (12.5 and 25%), at water temperatures of 23, 40, 55 and 70°C.

<u>Constant Strain Tests During Immersion</u>

The individual outcome of these tests is presented in Table 2 which shows the results for average % AF for 1 and 24 hours and 7 days immersion. Because of limitations on the number of specimens, not all 12 combinations were tested under all conditions. Table 3 contains the results of tensile tests on control specimens (tested dry at 23°C) and specimens subjected to standard conditioning (7 days immersion at room temperature unstrained), as well as those that survived the constant strain/water immersion tests.

TABLE 2—All individual results for constant strain tests. Average % AF for three replicates at 1 and 24 hours and 7 days immersion. Coefficient of variation is 25%.

| Temp. (°C) | Strain (%) | Exposure time | Combination no. 1 | 2 | 3 | 4 | 5 | 6 | 7 | 8 | 9 | 10 | 11 | 12 |
|---|---|---|---|---|---|---|---|---|---|---|---|---|---|---|
| 40 | 12.5 | 1 h | 0 | 0 | 0 | 0 | 20 | 0 | 0 | 0 | 10 | 70 | 40 | – |
| | | 24 h | 0 | 0 | 0 | 0 | 20 | 0 | 0 | 0 | 10 | 70 | 40 | – |
| | | 7 days | 0 | 0 | 0 | 0 | 80 | 0 | 0 | 0 | 40 | 90 | 70 | – |
| | 25 | 1 h | – | 0 | – | – | – | – | – | 30 | – | – | – | – |
| | | 24 h | – | 0 | – | – | – | – | – | 100 | – | – | – | – |
| | | 7 days | – | 0 | – | – | – | – | – | – | – | – | – | – |
| 55 | 0 | 1 h | 0 | 0 | 0 | 0 | 0 | 0 | 0 | 0 | 0 | 0 | 0 | – |
| | | 24 h | 0 | 0 | 0 | 0 | 0 | 0 | 0 | 0 | 0 | 0 | 0 | – |
| | | 7 days | 0 | 0 | 0 | 0 | 0 | 0 | 0 | 0 | 0 | 0 | 0 | – |
| | 12.5 | 1 h | 0 | 0 | 0 | 0 | 40 | 0 | 0 | 20 | 30 | 90 | 20 | – |
| | | 24 h | 0 | 0 | 0 | 0 | 40 | 0 | 0 | 20 | 30 | 90 | 20 | – |
| | | 7 days | 0 | 0 | 0 | 0 | 70 | 0 | 0 | 50 | 60 | 90 | 50 | – |
| | 25 | 1 h | 0 | 0 | 0 | 0 | 80 | 0 | 0 | 80 | 100 | 100 | 80 | 50 |
| | | 24 h | 0 | 0 | 0 | 0 | 80 | 0 | 0 | 80 | – | – | 80 | 50 |
| | | 7 days | 0 | 0 | 0 | 0 | 90 | 0 | 0 | 90 | – | – | 90 | 70 |
| 70 | 0 | 1 h | 0 | – | – | – | 0 | – | – | 0 | 0 | – | – | – |
| | | 24 h | 0 | – | – | – | 0 | – | – | 0 | 0 | – | – | – |
| | | 7 days | 0 | – | – | – | 0 | – | – | 0 | 0 | – | – | – |
| | 12.5 | 1 h | 0 | 0 | 0 | 0 | 10 | 0 | 0 | 50 | 0 | 100 | 40 | – |
| | | 24 h | 0 | 0 | 0 | 0 | 10 | 0 | 0 | 50 | 0 | – | 40 | – |
| | | 7 days | 0 | 0 | 0 | 0 | 40 | 0 | 0 | 90 | 50 | – | 70 | – |
| | 25 | 1 h | 0 | – | – | 0 | – | – | 0 | – | – | – | – | – |
| | | 24 h | 0 | – | – | 0 | – | – | 0 | – | – | – | – | – |
| | | 7 days | 0 | – | – | 0 | – | – | 0 | – | – | – | – | – |

TABLE 3—Results of tensile tests. A = control unconditioned specimens after standard curing. B = standard conditioning; specimens immersed at room temperature (RT) for 7 days without strain. C = specimens that survived the constant strain experiments as reported in Table 2 (T°C/Strain %).

| Combination | A | | B | | C | | | | | | | | | | | | | |
|---|---|---|---|---|---|---|---|---|---|---|---|---|---|---|---|---|---|---|
| | | | | | 40/12.5 | | 55/0 | | 55/12.5 | | 55/25 | | 70/0 | | 70/12.5 | | 70/25 | |
| | UTS[a] | % AF[b] | UTS | % AF | UTS | % AF | UTS | % AF | UTS | % AF | UTS | % AF | UTS | % AF | UTS | % AF | UTS | % AF |
| 1 | 0.79 | 0 | 0.76 | 0 | 0.68 | 0 | 0.68 | 0 | 0.67 | 0 | 0.63 | 0 | – | – | 0.27 | 0 | 0.38 | 0 |
| 2 | 0.78 | 0 | 0.70 | 0 | 0.63 | 0 | 0.68 | 0 | 0.65 | 0 | 0.60 | 30 | – | – | 0.40 | 30 | – | – |
| 3 | 0.75 | 0 | 0.83 | 10 | 0.70 | 0 | 0.65 | 0 | 0.65 | 0 | 0.60 | 0 | – | – | 0.57 | 10 | – | – |
| 4 | 0.76 | 0 | 0.72 | 0 | 0.76 | 0 | 0.72 | 0 | 0.68 | 0 | 0.63 | 0 | – | – | 0.60 | 0 | 0.55 | 0 |
| 5 | 0.83 | 0 | 0.77 | 0 | 0.28 | 40 | 0.67 | 0 | 0.40 | 30 | 0.18 | 80 | 0.52 | 35 | 0.38 | 20 | – | – |
| 6 | 1.02 | 0 | 0.69 | 80 | 0.75 | 10 | 0.72 | 40 | 0.67 | 20 | 0.65 | 60 | – | – | 0.35 | 100 | – | – |
| 7 | 0.78 | 0 | 0.79 | 0 | 0.78 | 0 | 0.84 | 0 | 0.73 | 0 | 0.70 | 0 | – | – | 0.40 | 0 | 0.38 | 15 |
| 8 | 0.94 | 0 | 0.88 | 0 | 0.63 | 10 | 0.75 | 0 | 0.40 | 50 | 0.25 | 80 | 0.25 | 80 | 0.15 | 90 | – | – |
| 9 | 0.85 | 0 | 0.85 | 0 | 0.55 | 50 | 0.65 | 0 | 0.40 | 40 | | *c* | 0.63 | 15 | 0.30 | 40 | – | – |
| 10 | 0.94 | 0 | 0.74 | 80 | 0.27 | 90 | 0.77 | 60 | 0.40 | 90 | | *c* | – | – | | *c* | – | – |
| 11 | 0.97 | 0 | 0.96 | 0 | 0.27 | 80 | 1.05 | 10 | 0.52 | 60 | 0.10 | 90 | – | – | 0.06 | 100 | – | – |
| 12 | 1.00 | 0 | 0.94 | 0 | 0.42 | 10 | 0.40 | 100 | 0.27 | 100 | 0.23 | 100 | – | – | 0.35 | 100 | – | – |

[a] Ultimate tensile strength (MPa) = peak load/bond area (tests in triplicate with a coefficient of variation of 10%).
[b] Coefficient of variation of 25%.
[c] 100% AF during immersion.

Figure 7 is a graphical representation of the combined results of average % AF across all 12 silicone/substrate pairs. Plots for 40°C/25%, 70°C/0% and 70°C/25% are not given due to the limited testing at these conditions.

It is also seen from Fig. 7 that tensile tests conducted immediately after exposure resulted in a higher level of adhesive failure for strained compared to unstrained specimens. In many cases failures were observed very early in the test, providing some potential for a more rapid assessment procedure than those currently recommended by US and European authorities.

Overall the level of discrimination of adhesive behaviour increases with the degree of severity of the test conditions. As seen from Table 3 tensile testing of unconditioned, i.e. dry samples (control), would provide no information about the relative adhesive strength of the 12 combinations. However as the harshness of the various parameters (strain, water temperature, immersion time) is increased, particularly strain, more and more combinations fail adhesively, thereby enabling the adhesive performance to be ranked. For example, combinations 1 and 4 which gave 0% AF at the most rigorous conditions (70°C/25%) were the best performers, while combinations 6 and 10 which gave 80% AF after conditioning at RT/0% would rank as very poor. It is also worth noting that the cohesive strength of the sealant (results for 0% AF) is generally reduced as the severity of the test conditions increases.

However, while being more useful than many of the current standardised tests, this technique does not yet provide objective measurements regarding the strength of adhesion. It is difficult to estimate the % AF during the test and separate tensile tests are necessary for specimens that do not fail during immersion, which appears to be the case for the majority of the sealant/substrate combinations. It would also be more desirable if the conditions were

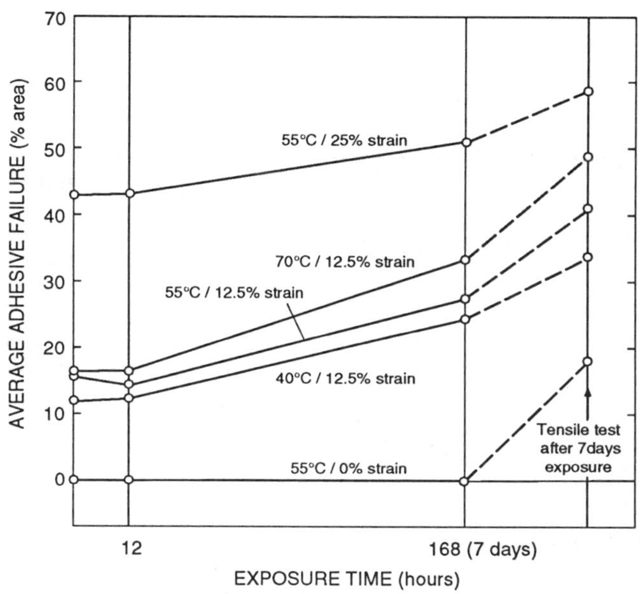


FIG. 7—Plot of average % AF for 12 sealant/substrate pairs versus immersion time for 5 different water temperature/constant strain combinations. The results of tensile tests on specimens that survived the 7-day exposure are also shown.

such as to always produce total joint failure during the test. However, in the case of constant initial strain tests, the onset of any failure results in stress relief which prevents the failure from propagating completely.

## Constant Stress Tests During Immersion

Preliminary assessment of the relative effects of the test parameters—In these dead loading experiments, three critical parameters are identified, namely stress, water temperature and immersion time. The relative effects of these parameters were investigated by testing combination 8 over a range of temperatures and stresses. This particular sealant/substrate pair was chosen because it had demonstrated intermediate adhesion levels during previous tests, and therefore should be useful in determining the relative effects of the above test parameters. For practical reasons 7 days was set as the maximum exposure time and tensile tests were conducted on surviving specimens. The results are given in Table 4. It should be noted that joint failures that occurred during dead loading were instant and catastrophic in all cases due to the absence of stress relief occurring in the "constant strain procedure" described earlier.

In general the incidence of failure during immersion increases with both increasing water temperature and load. As explained in a previous section, an increase of stress level in the presence of water has a significantly greater detrimental effect on the sealant/substrate bonds than an increase in temperature. These results also show that immersion at elevated

TABLE 4—Results of initial constant stress experiments to determine the relative effects of load and temperature (tests done in triplicate on combination 8). Control result = 0/0.94[a] (from Table 3).

| Temperature (°C) | Dead load stress (MPa) | | | |
|---|---|---|---|---|
| | 0 | 0.04 | 0.16 | 0.22 |
| 20 | – | – | 0/0.73[a] | 0/0.70[a] |
| 40 | – | 0/0.49[a] | 80/26<br>0/0.72[a] | 60/24<br>70/7<br>80/26 |
| 55 | 0/0.76[a] | 0/0.66[a] | 80/24<br>0/0.65[a] | 70/2<br>80/3<br>40/24 |
| 70 | 0/0.56[a] | – | 0/0.49[a] | – |
| 20 (without immersion) | – | – | – | 0/0.91[a] |

[a] These results represent the average percentage adhesive failure/ultimate tensile strength (MPa) for samples that were tested after surviving the 7-day exposure to the stated conditions. All other data represents the individual results, for samples that failed during immersion, as percentage adhesive failure/time to failure (h). Coefficient of variation for tensile tests is 10%.

temperature, even without stress, has a more severe effect on the cohesive strength of the sealant than constant stress alone, e.g. compare the tensile strengths for 70°C/0 MPa (0.56 MPa) and 20°C/0.22 MPa without immersion (0.91 MPa) with the dry control tensile strength (0.94 MPa). However, the combined effects of water temperature, immersion time and load are necessary to quickly produce adhesive failure. It is important to note that all specimens that survived the exposure period yielded 100% cohesive failure in subsequent tensile tests. Also in the two cases where only one out of three specimens failed during immersion (40°C/0.16 MPa, 55°C/0.16 MPa), the tensile strength of the two surviving specimens (sealant cohesive strength) was measured to be well above the applied dead load stress. This indicates that the failures during immersion occurred at stress levels well below the cohesive strength of the sealant, confirming that the failure must have initiated adhesively, i.e. at the sealant/substrate interface, although it may have propagated partly cohesively during the failure process.

As observed previously in the constant strain experiments during immersion, mechanical stress when applied above a certain threshold appears to degrade the adhesive bond more than elevated water temperature. For example, increasing water temperature at loads of 0.04 and 0.16 MPa did not significantly increase the level of failures. However, increasing the load for water temperatures of 40 and 55°C led to progressively higher frequencies of adhesive failure. The final important observation to be made is that the failures that occurred during immersion did so within about 24 hours.

<u>Selecting test conditions and procedure for constant stress tests</u>—On the basis of the above initial results, it was decided that the tests would be conducted at the one temperature of 50°C, and that the constant load would be increased in discrete steps until failure occurred within 24 hours. This critical load would constitute the test result and should discriminate the relative adhesive behaviour of the sealant/substrate system.

However, a technique for conducting the procedure had to be found that minimised the amount of testing required and yielded a statistically precise result. Simply increasing the load in steps and testing multiple samples at each step was considered unsatisfactory for various reasons. It would only identify the performance range of a particular system and not provide for more precise discrimination. Also to achieve accurate discrimination would require small steps and therefore an unacceptable amount of testing.

Therefore a particular statistical procedure known as the "Dixon up-and-down method for small samples" [6] was applied. This requires that initially the broad critical performance range be determined by conducting single specimen tests and increasing the load stepwise until failure occurs. A further sequence of tests, also on single specimens, is then performed within this critical range but using smaller load increments. On the basis of preliminary tests, and in order for the tests to be relevant to the structural use of sealants, the two stress step increments selected were 0.07 MPa (10 psi) and 0.014 MPa (2 psi). Fresh specimens were used for each test (i.e. specimens that did not fail were discarded). The details of the technique are described in Reference [5].

<u>Results of constant stress tests during immersion</u>—The results for all combinations, apart from 8, are given in Table 5. For comparison purposes the control, and control with standard conditioning from Table 3, are also provided. Sufficient specimens of combination 8 were not available to conduct these tests as these specimens were used for the initial trials reported in Table 4.

TABLE 5—Results of accelerated tests using variable dead loading and immersion in water at 50°C for 24 hours. A = control unconditioned specimens (from Table 3). B = specimens immersed at room temperature for 7 days without strain (from Table 3). C = accelerated constant load immersion tests.

| Combination | Sealant | A | | B | | C | |
|---|---|---|---|---|---|---|---|
| | | UTS[a] | % AF[b] | UTS | % AF | Failure stress during immersion (kPa) | % AF |
| 1 | 1 | 0.790 | 0 | 0.760 | 0 | 0.643[c] | 0 |
| 2 | 1 | 0.780 | 0 | 0.700 | 0 | >0.690[c] | 0 |
| 3 | 1 | 0.750 | 0 | 0.830 | 10 | 0.476 | 20 |
| 4 | 1 primed | 0.760 | 0 | 0.720 | 0 | 0.630[c] | 0 |
| 5 | 1 | 0.830 | 0 | 0.770 | 0 | 0.335 | 64 |
| 6 | 1 primed | 1.020 | 0 | 0.690 | 0 | 0.346 | 75 |
| 7 | 18 | 0.780 | 0 | 0.790 | 0 | 0.634[c] | 0 |
| 8 | 18 | 0.940 | 0 | 0.880 | 0 | – | – |
| 9 | 18 | 0.850 | 0 | 0.850 | 0 | 0.653[c] | 0 |
| 10 | 18 | 0.940 | 0 | 0.740 | 0 | 0.304 | 94 |
| 11 | 38 | 0.970 | 0 | 0.960 | 0 | 0.614[c] | 0 |
| 12 | 38 | 1.000 | 0 | 0.940 | 0 | 0.496 | 60 |
| 13 | 1 | 0.750 | 10 | 0.650 | 50 | 0.100 | 100 |
| 14 | 1 | 0.770 | 0 | 0.730 | 0 | 0.676[c] | 0 |
| 15 | 1 | 0.790 | 0 | 0.750 | 0 | 0.614 | 20 |
| 16 | 1 | 0.830 | 0 | 0.800 | 0 | 0.517 | 20 |
| 17 | 1 | 0.810 | 0 | 0.740 | 0 | 0.503 | 80 |
| 18 | 1 | 0.810 | 0 | 0.750 | 0 | 0.558 | 65 |
| 19 | 18 | 0.940 | 0 | 0.850 | 0 | 0.125 | 93 |
| 20 | 18 | 0.910 | 0 | 0.740 | 0 | 0.124 | 92 |
| 21 | 18 | 1.000 | 0 | 0.930 | 0 | 0.207 | 78 |
| 22 | 18 | 0.610 | 0 | 0.720 | 20 | 0.463 | 60 |
| 23 | 18 | 0.950 | 0 | 0.920 | 0 | 0.586[c] | 0 |
| 24 | 18 | 0.920 | 0 | 0.720 | 0 | 0.157 | 78 |

[a] Triplicate tests with coefficient of variation of 10%.
[b] Coefficient of variation of 25%.
[c] These results (0% AF) are indicative of the sealant cohesive strength under the conditions of this test.

As mentioned previously, straight control tests on unconditioned (dry) specimens would discriminate the adhesive behaviour of only combination 13. Standard conditioning (7 days immersion at room temperature) would further discriminate combinations 3 and 22. However, the current test would provide quantitative information on about 15 out of the 23 combinations tested. The failure stress during immersion in cases of 0% AF is a measure of the cohesive strength of the sealant under the particular test conditions. The results show that in all cases adhesive failure initiated at stresses well below the cohesive strength of the sealant, indicating that the adhesive bond across the sealant/substrate interface must have been the weakest part of the joint. Therefore the failure stress should be a reliable measure of relative adhesive behaviour. There is only a qualitative inverse correlation between stress at failure and % AF, and therefore the latter could not be used on its own to compare adhesive behaviour. Table 6 provides a ranking of all combinations that failed adhesively on the basis of this new test method.

TABLE 6—A ranking of the adhesive performance as given by the new accelerated test method[a].

| Combination description | Failure stress (MPa) |
|---|---|
| D1/1 | 0.100 |
| EA1/18 | 0.124 |
| EB3/18 | 0.125 |
| B9/18 | 0.157 |
| DB5/18 | 0.207 |
| D1/18 | 0.304 |
| P5/1 | 0.335 |
| M6/1 primed | 0.346 |
| P12/18 | 0.463 |
| M9/1 | 0.476 |
| G1/38 | 0.496 |
| EB3/1 | 0.503 |
| B9/1 | 0.517 |
| BA1/1 | 0.558 |
| M2/1 | 0.614 |

[a] Combinations B11/1, PBr/1, M9/1 primed, P1/18, G10/18, G10/38, DB5/1 and M2/18 all failed cohesively.

Primarily the combinations were chosen in order to cover as broad a range of materials (and therefore adhesive performance) as possible, as this was necessary for the development of the test. However, while the main aim of the experiment was not to examine the behaviour of the materials themselves, the following limited observations may be made:

(a) sealant 1 provides significantly better adhesion to aluminium anodised by the Chemel process than sealant 18;

(b) priming improves adhesion to green granite in the case of sealant 1;

(c) sealant 18 has better adhesion to substrate D1 (white fluorocarbon) than sealant 1; and

(d) sealant 38 has much better adhesion to reflective glass than to clear float glass.

The next phase in the development of this procedure will be to establish a minimum performance level for the conditions of this test. It is hoped that this will be achieved through correlations with traditional methods and the results of outdoor exposure trials, as well as actual in-service experience with common substrate/sealant systems.

However, it is possible to put the results into some practical perspective by noting that applying a minimum safety factor of say 3 (0.41 MPa or 60 psi using the common design criteria for SSG), combinations above M6/1 Primed in Table 6 would not comply.

Assessment of Test Sensitivity

In order to assess the sensitivity of the current test method, a set of sealant/substrate combinations were prepared using different cleaning solvents and primers. The results are given in Table 7.

TABLE 7—Assessment of test sensitivity. A = control unconditioned specimens. B = specimens conditioned by immersion at room temperature for 7 days without strain. C = current test results.

| Combin-ation | Substrate cleaning solvent | A | | B | | C | |
|---|---|---|---|---|---|---|---|
| | | UTS[a] | % AF[b] | UTS | % AF | Failure stress during immersion (MPa) | % AF |
| 19 | Ethanol | 0.940 | 0 | 0.850 | 0 | 0.125 | 93 |
| | Methyl ethyl ketone | 0.970 | 0 | 0.830 | 0 | 0.172 | 82 |
| | Ethanol & primer | 0.960 | 0 | 0.850 | 0 | 0.590 | 20 |
| 20 | Ethanol | 0.910 | 0 | 0.740 | 0 | 0.124 | 92 |
| | Methyl ethyl ketone | 0.930 | 0 | 0.880 | 0 | 0.185 | 80 |
| | Ethanol & primer | 0.950 | 0 | 0.830 | 0 | 0.550 | 10 |
| 24 | Ethanol | 0.920 | 0 | 0.720 | 0 | 0.157 | 78 |
| | Methyl ethyl ketone | 0.890 | 0 | 0.810 | 0 | 0.358 | 55 |
| | Ethanol & primer | 0.930 | 0 | 0.850 | 0 | 0.610 | 0 |

[a] Triplicate tests with coefficient of variation of 10%.
[b] Coefficient of variation of 10%.

The currently proposed test procedure is able to quantify the increase in bond strength due to the use of methyl ethyl ketone as the cleaning solvent in place of ethanol. The effectiveness of the primer is also well demonstrated. It is worth emphasising again that the usual test methods (A and B in Table 7) were not able to discriminate the effects of the various substrate treatment methods.

## SUMMARY AND CONCLUSIONS

Existing methods for assessing sealant adhesion have been shown to provide insufficient discrimination to be useful for ranking the performance of material combinations. The application of mechanical and environmental stress simultaneously has been found to degrade the adhesive interface specifically thereby providing a basis for discrimination.

Tests based on initial fixed strain and hot water immersion provided some distinction between the sealant/substrate combinations being tested. However, stress relaxation after the initiation of failure frequently stopped the progress of delamination, making it difficult to interpret the results.

The application of dead loads during hot water immersion proved to be the better combination of conditions. The technique developed involved varying the dead loading according to a particular statistical technique which gave a precise result for the minimum load required to produce adhesive failure. Using this procedure, the relative adhesive behaviour of 24 substrate/sealant combinations could be established. The technique was also used to discriminate the relative effectiveness of various substrate cleaning and preparation methods. Future work will attempt to set a minimum performance level through correlations with existing test methods, outdoor exposure trials and in-service experience of sealant/substrate systems.

## REFERENCES

[1] American Society for Testing and Materials (ASTM), "Standard Test Method for Adhesion-in-Peel of Elastomeric Joint Sealants," Designation: C794–80, 1980.

[2] Klosowski, J.M., *Sealants in Construction*, Marcel Dekker, Inc., New York, 1989.

[3] Iker, J. and Wolf, A.T., "Comparison of European and U.S. Testing/Qualification Procedures for Structural Glazing Silicone Sealants," *Proc. Symposium on Building Sealants: Materials Properties and Performance, ASTM STP 1069*, Thomas F. O'Connor, Ed., American Society for Testing and Materials, Philadelphia, 1990.

[4] Wu, S., *Polymer Interface and Adhesion*, Marcel Dekker, Inc., New York, 1982.

[5] Cerra, A.P., "Development of New Accelerated Adhesion Test for Sealants," *Journal of Testing and Evaluation*, in press.

[6] Dixon, W.J., "The Up-and-Down Method for Small Samples," *American Statistical Association Journal*, Vol. 69, 1965, p. 967.

Nick E. Shephard,[1] and James P. Wightman[2]

# AN ANALYSIS OF THE 180° PEEL TEST FOR MEASURING SEALANT ADHESION

**REFERENCE:** Shephard, N. E., and Wightman, J. P., **"An Analysis of the 180° Peel Test for Measuring Sealant Adhesion,"** Science and Technology of Building Seals, Sealants, Glazing and Weatherproofing: Fifth Volume, ASTM STP 1271, Michael A. Lacasse, Ed., American Society for Testing and Materials, 1996.

**ABSTRACT:** Sealant adhesion to different substrates is often assessed by using a 180° peel test. It is known that the peel force is a function of the viscoelastic response of the adhesive joint coupled with the intrinsic strength of the interphase. Measurement of the fracture energy as a function of straining rate can yield material properties by separating the bulk effects from the interphase strength. The results of this study highlight some important deficiencies of the ASTM peel test method. In particular, the fracture energy of a silicone sealant to glass and aluminum was measured at different rates, peel thicknesses and sealant moduli using the 180° peel test. It was determined that the sealant failed cohesively when tested on glass. For instances of cohesive failure, a plot of fracture energy vs. strain rate fit a power law model. The rate dependence of the fracture energy was shown to be proportional to the amount of uncrosslinked polymer in the sealant and this suggested an increase in dissipation. It was further shown that as the modulus of the sealant decreases the fracture energy increases. However, at low strain rates, the trend reverses and the high modulus sealant has better adhesion to glass than some of the low modulus sealants. Tests on specimens with aluminum substrates failed in accordance with specific test conditions: adhesive failure was more likely to occur when the strained thickness was small; when the strain rates were slow; or when the modulus was high. This study clearly demonstrated that peel testing at one rate and thickness can not adequately compare one sealant to another.

**KEYWORDS:** peel test, sealant, failure mode, silicone, viscoelastic, modulus, aluminum, glass

[1]Research Associate, National Science Foundation Science and Technology Center, Virginia Polytechnic Institute and State University, Blacksburg, VA 24061.

[2]Professor, Chemistry Department, Virginia Polytechnic Institute and State University, Blacksburg, VA 24061.

## INTRODUCTION

Sealants are used in numerous applications ranging from simple house hold caulking and weatherproofing, to expansion joints in roads and buildings, and to the more complicated load bearing structural glazing applications. The performance of the sealant is often assessed by measurement of 180° peel adhesion using standard test methods such as ASTM C 794-80. The common parameters measured are peel force and failure mode at one test speed and temperature. The failure mode is measured by visual inspection. The peel force is a function of the peeling rate, peeling thickness, temperature and strength of the interphase. An understanding of how these variables interact is necessary in order to translate the peel force results from one test configuration and speed to another. This paper reports on an experimental study to review the relationship between the peel force and failure mode as a function of sealant modulus, bond thickness, peeling rate and substrate type.

### Theoretical Considerations

ASTM Adhesion in peel--The 180 degree peel test, ASTM C 794-80 adhesion in peel, is the most common test currently used for studying sealant adhesion. It is also commonly used for lot to lot checking in the manufacturing plant and formulation comparison during product development. It is frequently called for in Federal Specifications and by the American Society of Testing Materials. The specifications usually call for a minimum peel force of 875N/m (5 pli) and 75% cohesive failure over the surface of the debonded area. Usually it is interpreted as a continuum mechanics test where the peel force and the failure mode are the only observations recorded. A water emersion cycle is often added to the cure cycle before testing to evaluate the durability of the joint. The failure mode is usually the most difficult aspect of the test to establish.

Fracture Energy Theory--The fracture mechanics of the peel test have been extensively studied and recently reviewed [1-8]. The energy needed to separate the two surfaces can be broken down as follows:

$$\mathcal{G} = \gamma + \phi_{\text{adhesive}} + \phi_{\text{bending}} + \phi_{\text{strain}} \tag{1}$$

where G is the strain energy release rate and $\gamma$ is the thermodynamic work of adhesion to separate the two surfaces. Dissipative energy expended in the bulk adhesive and in bending the stripping member as well as stored strain energy are included in the three $\phi$ terms. This equation is an extension of the Griffith fracture criterion and can be further generalized by the method of Andrews and Kinloch [9]. The three dissipative terms are combined in one function which is proportional to the work of adhesion between the two interfaces:

$$\mathcal{G} = \gamma\phi(\mathrm{T}, \dot{a}, \varepsilon) \tag{2}$$

where $\phi$ is a dissipation function dependent on the temperature (T), the crack speed ($\dot{a}$), and the average strain of the polymer ($\varepsilon$). The work of adhesion, $\gamma$, is not usually the same as Dupre work of adhesion. The value of $\gamma$ is often higher than expected for simple primary or secondary bonding between atoms in the joint because additional work is expended to strain the polymer chains to rupture. Therefore a new term called the intrinsic fracture energy, $\mathcal{G}_0$, is generally used. At constant temperature and strain, the following relationship holds:

$$\log \mathcal{G} = \log \mathcal{G}_o + \log \phi \qquad (3)$$

where $\phi$ is a function of strain rate only [9]. Equation 3 decouples the surface effects from the bulk effects. Changes measured as a function of surface changes will yield a series of curves which can be superimposed on each other by vertical shifting when plotted on a log-log graph of rate versus strain energy. The amount of vertical shifting equals:

$$\Delta \log \mathcal{G} = \log \mathcal{G}_{o1} + \log \phi_1 - \log \mathcal{G}_{o2} - \log \phi_2 \qquad (4)$$

where $\phi_1$ is equal to $\phi_2$. The above relationship has been useful for peeling conditions which do not include changes in the dissipation term (Fig. 1).

## EXPERIMENTAL METHOD

Materials

Sealant Formulation--A clear one-part acetoxy cured RTV-1 (room temperature vulcanized) silicone sealant was used in this work. The sealant type is commonly known as "acetoxy" or "RTV-1" which is an acronym for room temperature vulcanization one-part. This cure system was chosen for its good shelf stability and wide spread use in the construction industry. The cure chemistry of the RTV-1 silicone sealant is described by the following reactions [10]:

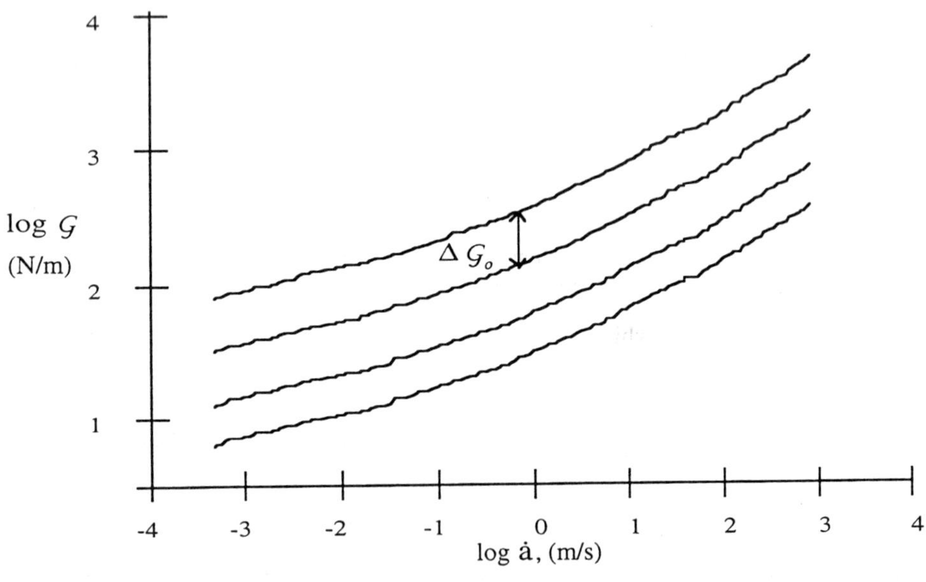


FIG. 1--Generalized example plot of fracture energy vs. crack growth rate (dissipation, $\phi$, is constant).

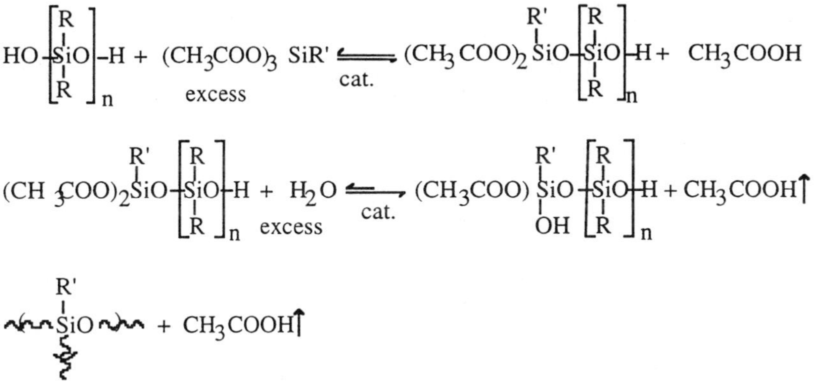

(i)

(ii)

(iii)

The catalyst used was dibutyltindilaurate and n is the number average molecular weight of the polymer. The letter R refers to the methyl end group and R' is a group having 50% methyl and 50% ethyl functionality. The first reaction occurs in the sealant package in the absence of water. The second reaction proceeds in the presence of water vapor after the sealant is applied to the adherend. Note the generation of acetic acid which is the source of the generic name, acetoxy sealant. The third reaction is a representation of the final network after the reaction has passed the gel point.

Four sealant formulations were prepared using the above cure chemistry. The formulations and sealant designations can be found in (Table 1). The primary ingredient was polydimethylsiloxane $\alpha$, $\omega$-diol with 42 000 g/mol number average molecular weight and with a viscosity of 14 000 centistoke. A methyl-terminated PDMS having 14 800 g/mol number average molecular weight (viscosity = 1,000 centistoke) was added to adjust the modulus of the sealant. The polydispersity of both polymers is approximately 2.2. The methyl-terminated polymer does not end link during the cure reaction. Instead, it acts like a diluent to reduce the cross link density. The diluent was added by replacing 0%, 16.7%, 33.4% and 50% of the reactive polymer respectively. The ratio of end-linker reactive sights to reactive polymer ends was also held constant at 16.5:1. Excess end-linker is used to reduce the cure rate and improve the shelf life of the uncured sealant. Fumed silica was added to the sealant to control the rheology and increase the modulus of the sealant [11]. The ratio of polymer, hydroxyl and methyl ended, to silica was held constant at 100:6. The sealants were prepared on a 3.7 L (1 gal.) double planetary mixer. The order of addition of the ingredients is as listed in Table 1 (diol.→filler). All mixing

TABLE 1--Sealant Formulations, parts by weight.

| Sealant Designation | 42,000 Mn HO∿∿OH | 1,000 Mn Me∿∿Me | end-linker* | Fumed Silica |
|---|---|---|---|---|
| A | 100.0 | 0.0 | 6 | 6 |
| B | 83.3 | 16.7 | 5 | 6 |
| C | 66.6 | 33.4 | 4 | 6 |
| D | 50.0 | 50.0 | 3 | 6 |

*8.26% catalyst in end-linker.

was performed under nitrogen atmosphere at 23°C. After mixing for twenty minutes, the sealant was degassed for 5 minutes under 800 mbar vacuum with mixing. The sealant was packaged into 178 ml (6 oz.) cartridges and centrifuged to remove any air entrapped during packaging. The yield was approximately 10 cartridges for each formulation.

Glass--Single strength float glass was obtained from the local hardware store and cut into pieces 8 cm x 16 cm x 0.23 cm. Each sample was cleaned with methyl ethyl ketone solvent three times and allowed to air dry before preparing the joint. The high tin side of the glass was used for adhesion measurements.

Aluminum--An 6061 T6 aluminum alloy was cleaned and cut in the same manner as the glass. The native oxide was not altered before testing.

## Methods

Modulus Measurement--The modulus of the sealant was measured using an Instron testing machine Model No. 1123 with a 500 g load cell. Dogbone shaped test specimens were cut from sheets of sealant using an ASTM D-638-5 carbon steel die. A strain gauge was used to measure the change in length of the 25.4 mm (1 in.) gauge length. The cross head speed was 8.5 mm/s (20 in./min.) resulting in a strain rate of 0.33 $mm \cdot mm^{-1}/s$.

Peel Testing--A modified version of the ASTM C 794-80 adhesion in peel test was employed. The test specimens were prepared by tooling a specified thickness of sealant onto the surface of the cleaned The thickness was controlled by using draw down bars of the correct thickness. Immediately after the sealant was applied a fiberglass cloth was pressed onto the sealant. Before the cloth was applied, sealant was tooled into both sides of the cloth. Finally, a second layer of sealant was applied to the top of the cloth to give the final thickness of the specimen, t, which was kept constant at 6.4 mm (0.25 in.) (Fig. 2). The strained thickness, b, of the sealant was varied over three different thicknesses ranging from 0.5, 3.2 and 6.4 mm (0.02, 0.125, 0.25 in.). This controlled the thickness of the region in which failure could occur. The peel cloth used was 213 $g/m^2$ (9 oz./sq.yd.) fiberglass cloth with a silane sizing agent. The peel width was 12.7 mm (0.5 in.) and was controlled by cutting the peel samples after full cure using a razor and a straight edge. The sealant was allowed to cure for 21 days at ambient conditions before testing. The strain in the tearing member is negligible because the fiberglass cloth is much higher modulus than the sealant. The bending moment is also negligible for this system of adherend and adhesive because the bending forces are much smaller than the forces needed to peel the joint apart. The strain energy release rate can be approximated by the following equation [12].

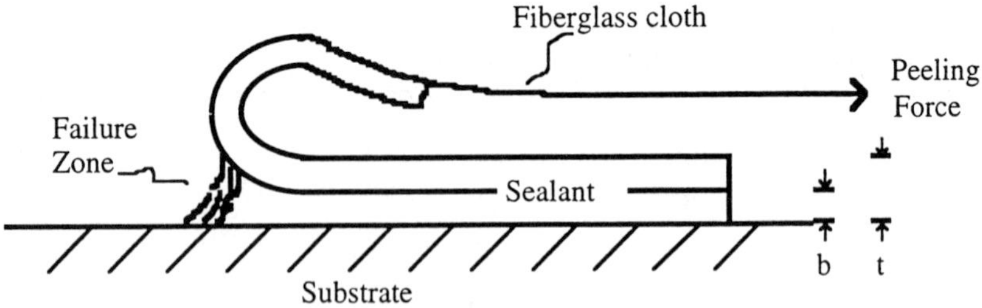


FIG. 2--Peel test geometry. Peeling width, 0.0127 m.

$$\mathcal{G} = \frac{P}{w}(1 - \cos\theta) \quad (5)$$

where P is the peeling force, w the peel width and θ the peeling angle.

The samples were tested using an Instron testing machine Model No. 1123 with a variable range 90 kg load cell. The samples were clamped into the Instron with an aluminum fixture designed to prevent breaking of the glass substrate. A precrack was made in the sealant by using a razor blade to cut through the sealant next to the substrate. The fiberglass cloth was pulled away from the substrate at a peeling angle of 180°. The peeling rate was varied over the following range, 0.085, 0.85, 8.5 mm/s (0.2, 2.0, 20.0 in./min). Continuous data acquisition of the peeling force was achieved through the computer interface. The average peeling force at steady state peeling was used in all the data analysis. Visual inspection was used to determine the location of the crack growth.

## RESULTS AND DISCUSSION

Sealant Modulus

The stress-strain isotherms were analyzed by plotting the engineering stress vs. the engineering strain (Fig. 3). The energy to break can be calculated by taking the integral of the stress-strain curve:

$$W_b = \frac{W}{A_o l_o} = \int_0^{\varepsilon_b} \sigma d\varepsilon \quad (6)$$

where $W_b$ is the energy to break, $A_o$ the initial area and $l_o$ the gauge length, ε the strain ($\varepsilon = (l - l_o)/l_o$ where l is the new length of the stretched neck), σ the stress ( $F/A_o$, F= force). Shown in (Fig. 3) are solid lines representative of the slope of the line tangent to

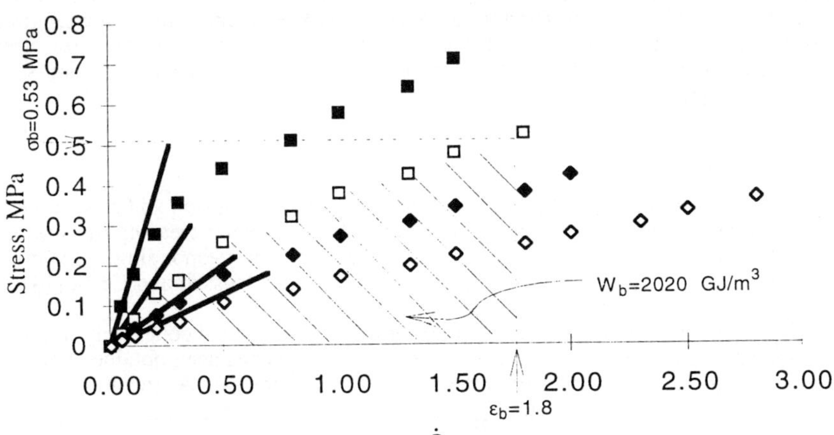


FIG. 3--Stress vs. strain data at 0.33/s strain rate. Sealant materials: ■ A; □ B; ◆ C; ◇ D.

TABLE 2 -- Stress-strain data for different sealant formulations subjected to tensile elongation.

| Sealant Designation | $\varepsilon_\beta$, Strain to Break | $\sigma_b$, Stress to Break, MPa | E, Modulus, MPa | $W_b$, Energy to Break, $GJ/m^3$ |
|---|---|---|---|---|
| A | 1.5 | 0.71 | 1.92 | 1250 |
| B | 1.8 | 0.53 | 0.84 | 2020 |
| C | 2 | 0.47 | 0.4 | 1910 |
| D | 2.8 | 0.37 | 0.26 | 1910 |

the origin which yields the engineering modulus.

As provided in (Table 2.), the strain to break, $\varepsilon_b$, increases with increasing diluent level (i.e. from A→D); whereas, the stress to break, $\sigma_b$, and the modulus, E, are both reduced by adding diluent. The energy to break, $W_b$, when diluent is present is nearly twice that of the sealant without diluent for example, the value of $W_b$ for sealants A and D are 1250 $GJ/m^3$ and 1910 $GJ/m^3$ respectively.

Peel Adhesion

The independent variables for the peel test study were peeling rate, fiberglass cloth embedded depth, sealant modulus and substrate type. The dependent variables studied were peel force and failure mode. The strain energy release rate, $\mathcal{G}$, was calculated according to equation 5 and the results provided in Table 3 are average values based on 3 pulls of each sample condition. A log-log plot of the strain energy release rate against the peeling rate resulted in straight lines for each thickness and modulus condition. The effect of thickness can be corrected by plotting the strain rate instead of the peeling rate and can be estimated by the following equation [13]:

$$\dot{\varepsilon} = \frac{\dot{a}}{b} \quad (7)$$

where $\dot{\varepsilon}$ is the strain rate and $\dot{a}$ is the peeling rate. This is analogous to measuring strain rate instead of cross head speed for a uniaxial tension experiment. The gauge length is assumed to be the same as the strained thickness, b. At constant peeling rate, the strain rate increases as the strained thickness, b, decreases. A linear fit of the data results in the following equation:

$$\mathcal{G} = k\dot{\varepsilon}^n \quad (8)$$

where k is a constant related to the thermodynamic work of adhesion and n is a constant related to rate dependent dissipation. It can readily be seen that increasing the strained thickness, b, will result in a lower effective strain rate and a correspondingly lower strain energy release rate as was previously reported by Gutowski, Russell and Cerra [14].

Glass--All the specimens, with the exception of one, failed cohesively in the strained sealant region (Fig. 4). A reasonable fit to the data has been obtained with the coefficient of variance, $r^2$, greater than 0.8 for all cases (Table 4). As the amount of diluent in the polymer increases, the exponent n increases (i.e. from A→D). The correspondence between the coefficient n and the diluent level can be readily seen in (Fig. 5) The amount of non-reacted polymer in Fig. 5 is increased by 5% to account

TABLE 3 -- 180° Peel Adhesion Data for All Test Conditions.

| Peeling Speed (mm/s) | Sealant Modulus (MPa) | b, Constrained Thickness (mm) | Peel Force Aluminum Substrate (N/m) | Failure Mode* Aluminum Substrate | Peel Force Glass Substrate (N/m) | Failure Mode* Glass Substrate |
|---|---|---|---|---|---|---|
| 0.084 | 0.10 | 0.50 | 35 | 0.0 | 525 | 1.0 |
| 0.084 | 0.10 | 3.20 | 35 | 0.0 | 385 | 1.0 |
| 0.084 | 0.10 | 3.20 | 35 | 0.0 | 350 | 1.0 |
| 0.084 | 0.10 | 6.40 | 245 | 1.0 | 350 | 1.0 |
| 0.084 | 0.18 | 0.50 | 140 | 0.0 | 630 | 1.0 |
| 0.084 | 0.18 | 3.20 | 105 | 0.0 | 525 | 1.0 |
| 0.084 | 0.18 | 6.40 | 350 | 0.0 | 350 | 1.0 |
| 0.084 | 0.25 | 0.50 | 105 | 0.0 | 350 | 0.5 |
| 0.084 | 0.25 | 3.20 | 53 | 0.0 | 350 | 1.0 |
| 0.084 | 0.25 | 6.40 | 350 | 0.0 | 350 | 1.0 |
| 0.084 | 0.37 | 0.50 | 53 | 0.0 | 245 | 0.0 |
| 0.084 | 0.37 | 3.20 | 70 | 0.0 | 525 | 1.0 |
| 0.084 | 0.37 | 6.40 | 385 | 0.0 | 560 | 1.0 |
| 0.084 | 0.37 | 6.40 | 352 | 0.0 | 509 | 1.0 |
| 0.84 | 0.10 | 0.50 | 280 | 0.5 | 771 | 1.0 |
| 0.84 | 0.10 | 3.20 | 350 | 0.5 | 630 | 1.0 |
| 0.84 | 0.10 | 6.40 | 701 | 1.0 | 630 | 1.0 |
| 0.84 | 0.18 | 0.50 | 210 | 0.0 | 736 | 1.0 |
| 0.84 | 0.18 | 3.20 | 53 | 0.0 | 490 | 1.0 |
| 0.84 | 0.18 | 6.40 | 175 | 0.0 | 525 | 1.0 |
| 0.84 | 0.25 | 0.50 | 70 | 0.0 | 560 | 1.0 |
| 0.84 | 0.25 | 3.20 | 35 | 0.0 | 490 | 1.0 |
| 0.84 | 0.25 | 6.40 | 455 | 1.0 | 455 | 1.0 |
| 0.84 | 0.37 | 0.50 | 70 | 0.0 | 420 | 1.0 |
| 0.84 | 0.37 | 3.20 | 53 | 0.0 | 455 | 1.0 |
| 0.84 | 0.37 | 6.40 | 455 | 1.0 | 630 | 1.0 |
| 8.4 | 0.10 | 0.50 | 1751 | 1.0 | 1541 | 1.0 |
| 8.4 | 0.10 | 3.20 | 1576 | 1.0 | 1191 | 1.0 |
| 8.4 | 0.10 | 6.40 | 1296 | 1.0 | 1086 | 1.0 |
| 8.4 | 0.10 | 6.40 | 1136 | 1.0 | 1289 | 1.0 |
| 8.4 | 0.18 | 0.50 | 701 | 0.5 | 1261 | 1.0 |
| 8.4 | 0.18 | 3.20 | 806 | 0.5 | 701 | 1.0 |
| 8.4 | 0.18 | 6.40 | 841 | 1.0 | 806 | 1.0 |
| 8.4 | 0.25 | 0.50 | 280 | 0.5 | 876 | 1.0 |
| 8.4 | 0.25 | 3.20 | 175 | 0.0 | 665 | 1.0 |
| 8.4 | 0.25 | 6.40 | 701 | 1.0 | 701 | 1.0 |
| 8.4 | 0.37 | 0.50 | 140 | 0.0 | 665 | 1.0 |
| 8.4 | 0.37 | 0.50 | 175 | 0.0 | 771 | 1.0 |
| 8.4 | 0.37 | 3.20 | 210 | 0.5 | 560 | 1.0 |
| 8.4 | 0.37 | 6.40 | 560 | 1.0 | 736 | 1.0 |

* A rating system of 0 for interfacial failure and 1 for cohesive failure was used. Intermediate values indicate mixed mode failure.

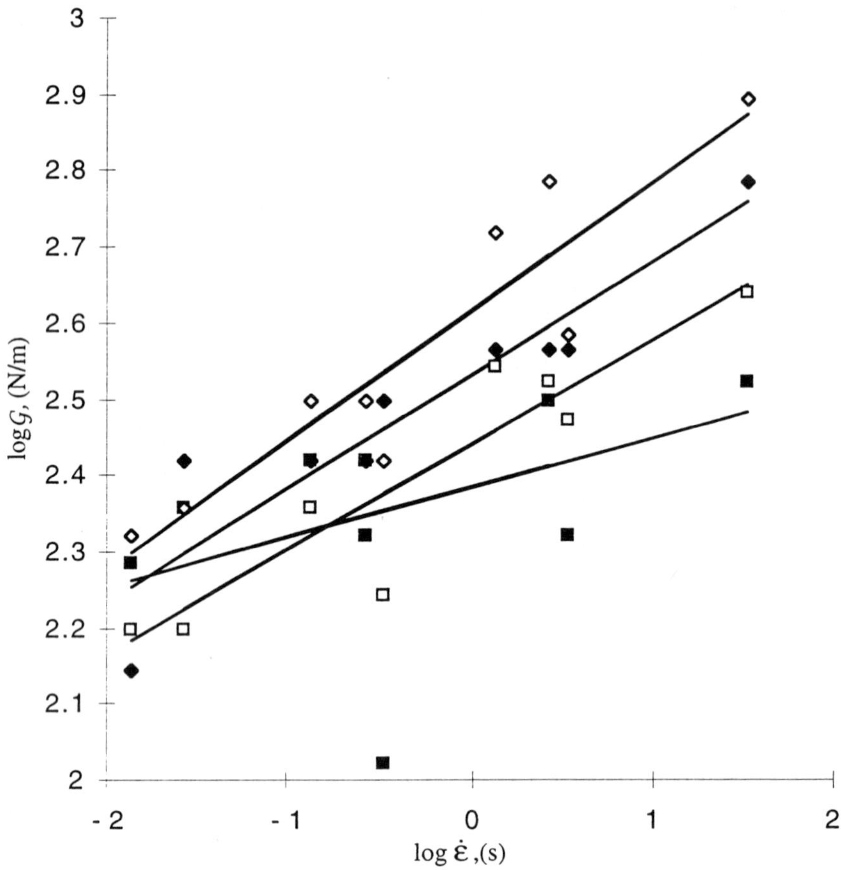


FIG. 4--Fracture energy vs. strain rate for the glass substrate. ■ A. □ B; ◆ C; ◇ D.

TABLE 4 -- Regression data for glass adhesion.

| Sealant Designation | E, Modulus, MPa | $W_b$, Energy to Break, GJ/m$^3$ | k | n |
|---|---|---|---|---|
| A | 1.92 | 1250 | 243(17) | 0.0651(78) |
| B | 0.84 | 2220 | 277(7) | 0.138(16) |
| C | 0.40 | 1910 | 341(7) | 0.149(15) |
| D | 0.26 | 1910 | 415(8) | 0.171(16) |

*numbers in parenthesis are % standard deviation.

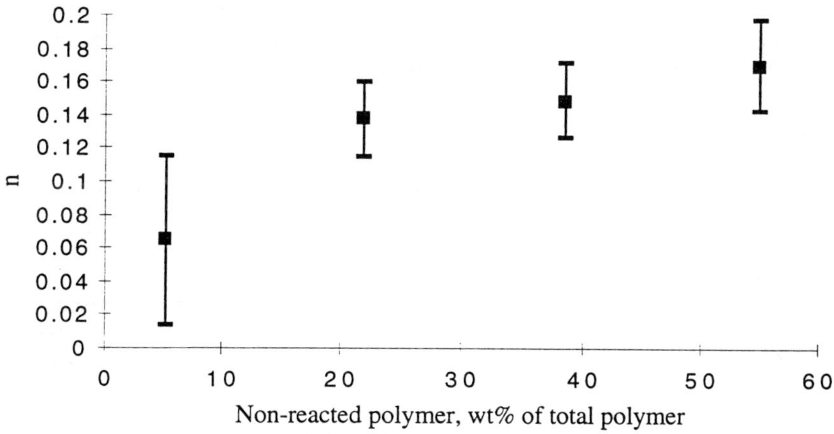


FIG. 5--Correspondence of the coefficient n with non-reacted polymer for glass fracture data. 5% is added to the amount of non-reacted polymer to account for cyclics in the polymer.

for the equilibrium level of non-reactive cyclic polymer chains which occur during the manufacture of the hydroxyl terminated polydimethylsiloxane. This suggests that a greater amount of dissipation occurs in the sealants which have more unreacted polymer in the network. The constant n can also be loosely correlated to the energy to break. This is consistent with the fact that the energy to break and the constant n are both related to the amount of dissipation which is occurring during the experiment. The fact that the level of dissipation is increasing precludes the use of equation 4. The constant k also increases with decreasing modulus. However, this may not indicate an increase in the threshold or intrinsic fracture energy. In fact, the fracture energy for the high modulus sealant is greater at low strain rates than the fracture energy for the lower modulus sealant. This suggests that the intrinsic fracture energy of the high modulus sealant may be greater than that of the low modulus sealants. The high rate data suggests the opposite trend. This is an excellent example of the problems associated with measuring fracture energy at one rate.

Aluminum--The data obtained from test specimens adhered to aluminum substrates is difficult to interpret. Peel strength is strongly influenced by the failure mode. For the aluminum adherend, the failure mode changes between adhesive and cohesive for various sample conditions within the experiment. For purposes of comparison, the aluminum data was analyzed according to equation 8 The coefficient of variance, $r^2$, for the regression analysis was less than 0.5 for all cases possibly due to events which invalidate equation 8. The dissipation due to straining of the sealant will be lower when the failure mode is adhesive. This occurs because the crack propagates before the sealant is strained to break. Therefore, the dissipative term in equation 8 is no longer a smooth function of the strain rate. However, the highest strain energy release rates can be correlated directly with cohesive failure. In addition, the trends of modulus and strain rate were similar to that of the glass data.

Cohesive failure occurs more frequently when the sealant modulus is decreased, the peeling rate is increased or when the peeling thickness is increased (Fig. 6). The same trends are also seen for the glass data (Fig. 6). However, the total occurrence of cohesive failure is much higher for the glass substrate indicating better adhesion at the interface of the glass substrate than at the interface of the aluminum substrate.

## CONCLUSION

The adhesion of a model sealant formulation to glass and aluminum has been tested using the 180° peel test. The glass test specimens failed cohesively through the sealant for all but one condition. The fracture energy has been related to the test geometry, peeling rate and sealant modulus. For instances of cohesive failure, a power law model for fracture energy applies. The peeling rate of the crack is not used. Instead, the strain rate of the sealant is used to correct for the changes in sealant thickness. As the modulus of the sealant decreases the rate dependence increases. This suggests an

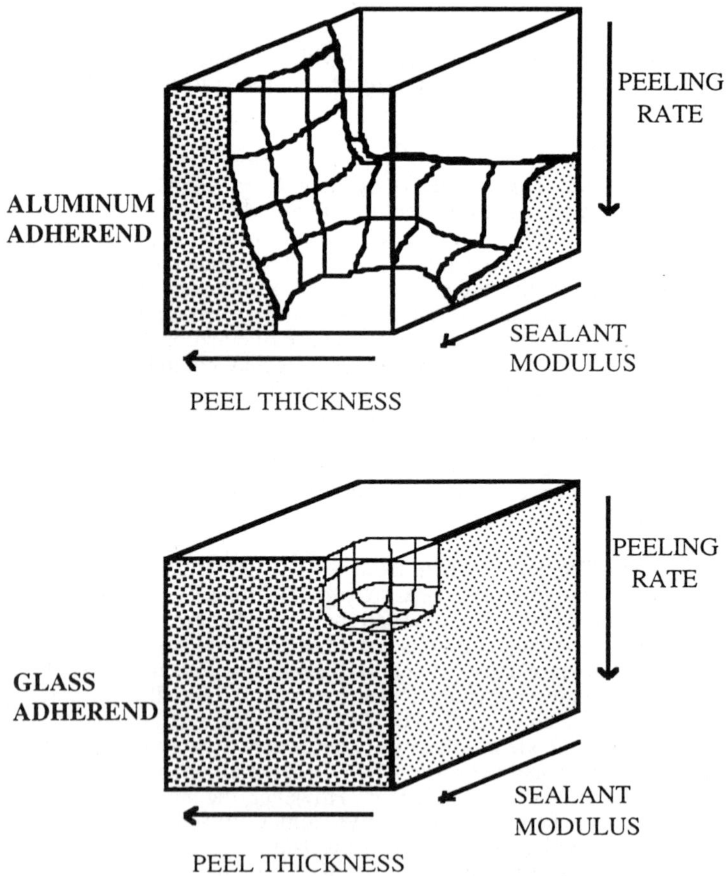


FIG. 6--Region of cohesive failure for glass and aluminum.

increase in the amount of dissipation in the sealant as the amount of unreacted polymer is increased. In general, the fracture energy increases for the sealants with higher levels of dissipation. However, fracture energy measurements taken at lower strain rates suggest that the higher modulus sealant may have better adhesion than the lower modulus sealant.

The aluminum test specimens failed adhesively or cohesively depending on the test conditions. Cohesive failure yields a high fracture energy. Cohesive failure occurs more frequently when the sealant modulus is decreased, the peeling rate is increased or when the peeling thickness is increased. Adhesive failure results when the stress is concentrated near the interphase. This occurs when the strained thickness is small. In addition, slow strain rates allow the sealant to relax and transfer the stress to the interphase. The glass data follows the same trends. However, the total occurrence of cohesive failure is much higher for the glass substrate indicating better adhesion at the interface of the glass substrate than at the interface of the aluminum substrate.

The above results highlight some important deficiencies in the ASTM peel test method. Peel testing at one rate and thickness can not adequately compare one sealant to another. The peel force is a function of the viscoelastic response of the adhesive joint coupled with the intrinsic strength of the interphase. Measurement of the fracture energy as a function of straining rate can yield material properties by separating the bulk effects from the interphase strength.

## ACKNOWLEDGMENTS

This research was funded by the Dow Corning Corporation at Midland, Michigan and the Center for Adhesive and Sealant Science at Virginia Tech.

## REFERENCES

[1] D. H. Kaelble, Journal of Colloidal Science, Vol. 19, 1964, p. 413.

[2] A. N. Gent and A. J. Kinloch, Journal of Polymer Science, A-2, Vol. 9, 1971, p. 659.

[3] A. N. Gent and J. Schultz, Journal of Adhesion, Vol. 3, 1972, p. 281.

[4] A. N. Gent and G. R. Hamed, Rubber Chemistry and Technology, Vol. 51, 1978, p. 354.

[5] D. Maugis, Journal of Adhesion, Vol. 1, No. 2, 1987, p. 105.

[6] T. Igarshi, "Peel-Strength and Energy Dissipation," in: Adhesive Joints: Formation, Characteristics and Testing, K. L. Mittal, Ed. Plennum Press, New York, 1984, p. 419.

[7] J. R. Huntsberger, "Interpreting Peel Tests," in: Treatise on Adhesion and Adhesives, R. L. Patrick, Ed. Vol. 6, Marcel Dekker, New York, 1989, p. 1.

[8] G. R. Hamed, "Energy Conservation During Peel Testing," in: Treatise on Adhesion and Adhesives, R. L. Patrick, Ed. Vol. 6, Marcel Dekker, New York, 1989, p. 33.

[9] E. H. Andrews and A. J. Kinloch, Proceedings of the Royal Society (London), Vol. A332, 1973, p. 401.

[10] W. Noll, Chemistry and Technology of Silicones, Academic Press, 1968.

[11] H. Cochrane and C. S. Lin, Rubber Chemistry and Technology, Vol. 66, 1993, p. 48.

[12] T. Hata, *Kobunshi Kagaku,* **4**, 1947, p. 67.

[13] A. N. Gent and R. P. Petrich, Proceedings of the Royal Society, Vol. A310, 1969, p. 433.

[14] W. S. Gutowski, L. Russell, and A. Cerra, "New Tests for Adhesion of Silicone Sealants," Science and Technology of Building Seals, Sealants, Glazing and Waterproofing: Second Volume, ASTM STP 1142, ed. J. M. Klosowski, American Society for Testing and Materials, Philadelphia, Pennsylvania, 1992.

Philippe J. Hubin-Eschger[1]

# IMPROVEMENT OF INITIAL ADHESION TO ALUMINUM OF INSULATING GLASS POLYURETHANE SEALANTS USING A TERTIARY AMINE BASED CURING SYSTEM

---

**REFERENCE:** Hubin-Eschger, P.J., **"Improvement of Initial Adhesion to Aluminum of Insulating Glass Polyurethane Sealants Using a Tertiary Amine Based Curing System"**, Science and Technology of Building Seals, Sealants, Glazing, and Waterproofing: Fifth Volume ASTM STP 1271, Michael A. Lacasse, Editor, American Society for Testing and Materials, 1996.

**ABSTRACT:** Hydroxylated polybutadiene based insulated glass sealants are well known to have a very low moisture vapor transmission rate and an excellent adhesion to glass and aluminum. Since the beginning of their industrial development, the best and most known and effective curing system for these sealants has been an organomercuric ester based catalyst.
The current and future pressure regarding ecological and toxicological issues will make this type of product increasingly difficult to use. This paper presents the results of work to develop other catalytic systems, which were obtained with the tertiary amine dimethylbenzylamine. The final improvement of initial adhesion to aluminum is obtained with the use of phenolic derivatives of coumarone indene.

**KEYWORDS** : insulating glass sealant, hydroxylated polybutadiene, dimethylbenzylamine, monohydroxylated coumarone indene, initial adhesion.

---

## INTRODUCTION

Liquid hydroxylated polybutadiene (HTPB) based insulating glass sealants have been in use for more than fifteen years. They are well known for their long term performance and particularly for having a very low moisture vapor transmission rate (MVTR) and an excellent adhesion to glass and aluminum.

Since the beginning of their industrial development, to achieve the optimum ultimate properties, these sealants have used a curing system catalyzed with a mercuric salt, namely the phenyl mercury ester (PME) of a C10 monobasic acid.

---

[1] Technical Manager for Poly bd® Resins at Elf Atochem, Groupement de Recherches de Lacq, 64170 Artix, FRANCE

Nowadays, concerns with ecological and toxicological effect of these products have placed them in question. In some countries such as the United States of America, their use for IGS insulating glass sealant catalysis has already been banned. In fact, in many countries serious limitations were imposed and the threat of discontinuance of manufacturing is real.

It was therefore necessary to look for other catalysts that could possibly provide similar results. In this paper, we shall present the work done in our Application Laboratory of LACQ (France) in contribution to this aim.

An insulating glass unit is designed to reduce the thermal exchange between the two sides of a window. This is obtained, in principle, by providing a layer of anhydrous air, or other gas, between two or more panes of glass.

To achieve these requirements, a conventional "double seal" unit is made up of the following elements :

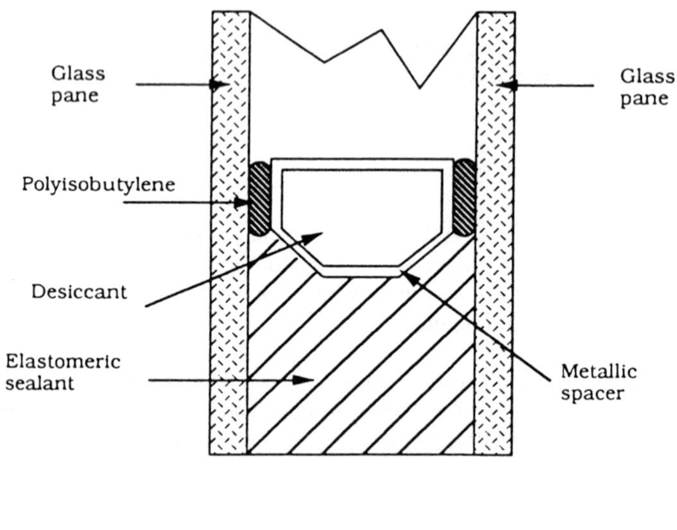


FIG. 1--Schematic of a conventional "double seal" insulating glass unit (IGU).

- Two panes of glass ;
- Hollow metallic spacer, in most cases aluminum, filled with a dessicant ;
- Tape of polyisobutylene between both sides of the spacer and the internal sides of the panes ;
- Elastomeric sealant surrounding the entire perimeter of the unit. It insures impermeability to water vapor and provides elasticity to absorb extension and contraction due to the temperature changes.

To fulfil an adequate level of performance, the sealants must therefore have a low MVTR (e.g. 1.0 $g/m^2/day$) and adhere perfectly to both glass and aluminum.

In addition to that, they must cure rapidly thus providing ease of handling and remain stable during the intended service life.

## IGS FORMULATION

A typical formulation for a two part HTPB based IGS used as a starting point is shown in Table 1.

TABLE 1 -- Formulation for a two-part HTPB based IGS using PME

| Ingredients | Parts by weight |
|---|---|
| Part A | |
| HTPB | 100 |
| Phthalate Plasticizer | 90 |
| Calcium carbonate 1 (a) | 120 |
| Calcium carbonate 2 (b) | 240 |
| Fumed silica | 5 |
| Titanium dioxide | 15 |
| Weathering agents | 3 |
| PME | 0.25 |

(a) Coated ; average particle size 0.04 μm
(b) Coated ; average particle size 0.45 μm

| Ingredients | Parts by weight |
|---|---|
| Part B | |
| MDI (c) | 100 |
| Phthalate plasticizer | 160 |
| Silane adhesion promotor | 48 |
| Carbon black | 38 |
| Molecular sieves | 24 |

(c) Diphenylmethane 4,4' diisocyanate

A mix ratio of 10/1 by volume was used that made the IGS convenient for automatic machine distribution.

When trying to substitute the phenyl mercury ester for another catalyst, one faces a triple challenge ; the IGS must :

- Have a long pot life followed by a rapid cure which insures that the tack-free time is minimized thus permitting handling and shipping insulating glass unit as quickly as possible ;

- Encourage the less favorable hydroxyl-isocyanate reaction in preference to the water-isocyanate reaction ;

- Reach quickly an adequate level of initial adhesion to the aluminum spacer ; adhesion to glass and long term adhesion to aluminum are usually never a problem.

Of the large number of catalysts that were evaluated, mostly metallic salts and amines and their mixtures, the very best overall results were obtained with the tertiary amine dimethylbenzylamine (DMBA), typically used in the polyurethane foam industry.

A review of how the parameters of the formulation were adjusted to reach a satisfactory compromise is first discussed and thereafter a comparison is made between the performance results using the new formulation in relation to the conventional system. Then we shall detail the comparative figures obtained after all the changes.

1 - Curing characteristics

A simple increase of the total amount of catalyst in the formulation gave a very similar profile of cure : 0.55 part by weight of DMBA provided similar curing characterics that of an IGS formulation with 0.25 part by weight of PME.

2 - Influence of water

One of the main strengths of PME is its ability to favor the reaction of isocyanate groups with the hydroxyl groups of polyols rather than with water. In the formulation water is brought mainly by the calcium carbonates. Its level is regulated with the molecular sieves. When switching to DMBA, it is necessary to decrease the final quantity of water left in the formulation by increasing the amount of molecular sieves. However water is useful for other properties and its reduction affects primarily the rheological behavior of the sealant that has to be non-sag after the mixing of its two parts. Two solutions to the problem were tried : Increase the amount of fumed silica or change the ratio between the two calcium carbonates. The second one was chosen because of its better reproducibility.

3 - Initial adhesion to aluminum

As previously stated, the change of catalyst often has a negative effect on this parameter. This is the case when using DMBA. Not only the peel strength is lower but the rupture becomes adhesive on aluminum. After many trials with all kinds of additives it was concluded that a few high molecular weight monohydroxylated compounds were very efficient in restoring this property and even enhancing it. Among those, the best results were obtained with some complex mixtures of phenolic copolymer hydrocarbons known under the specific name of monohydroxylated coumarone indene derivatives.

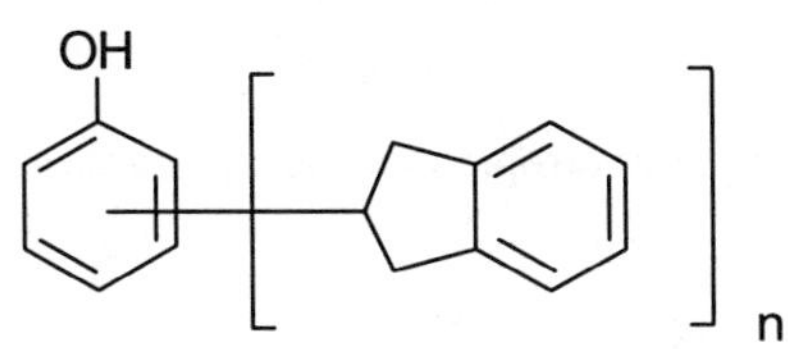


FIG. 2--Formula of a monohydroxylated coumarone indene derivative

Two grades, differing by their OH-content are commercially available. They have very little effect on other characteristics of the sealant and their cost is low.

The new formulation is given in Table 2 :

TABLE 2 --Formulation for a two-part HTPB based IGS using DMBA.

| Ingredient | Parts by weight |
|---|---|
| Part A | |
| HTPB | 100 |
| Phthalate plasticizer | 86 |
| Calcium carbonate 1 | 160 |
| Calcium carbonate 2 | 195 |
| Fumed silica | 5 |
| Titanium dioxide | 15 |
| Weathering agents | 3 |
| Molecular sieves | 5 |
| Monohydroxylated compound (a) | 4 |
| Dimethylbenzylamine | 0.55 |

(a) Coumarone indene derivative, OH-content % : 2.5-2.7

The components of part B are strictly identical to those provided in Table 1 as well the mix ratio remains the same.

A comparison of the performances between the two formulations is made.

## PERFORMANCE PROPERTIES

1 - Rheological properties

As shown in Table 3 viscosity, thixotropic index and sag are essentially unchanged.

TABLE 3 --Rheological properties

| Catalyst | Mix viscosity, Pa.s | Thixotropic index | Sag, mm |
|---|---|---|---|
| PME | 750 (a) | 1.65 | 1.0-1.5 |
| DMBA | 730 (a) | 1.62 | 1.5-2.0 |

(a) RHEO Viscosimeter. Model HV8/Cylinder 11/2.5 rpm/25°C

Thixotropic index is the ratio between the viscosities at 2.5 and 5 rpm. Sag resistance is determined with a Boeing jig according to ASTM D 2202 Test Method for Slump of Sealants. An insulating glass sealant is usually permitted to slump up to 3 mm.

2 - Curing characteristics

The pot life is estimated by the weight of material extruded under standard conditions (DIN 52456) over a certain period of time and the Shore A hardness is measured according to ASTM D 2240 Test Method for Rubber Property - Durometer Hardness.

TABLE 4 -- Curing characteristics

| Catalyst | Pot life, min | Hardness, Shore A | | | | | | Tack free |
|---|---|---|---|---|---|---|---|---|
| | | 3h | 4h | 5h | 6h | 24h | 1 day | time, h |
| PME | 53 | 28 | 32 | 34 | 36 | 42 | 42 | 3.0 |
| DMBA | 49 | 25 | 30 | 33 | 36 | 41 | 42 | 3.5 |

As can be seen from the results provided in Table 4 either of the IGS formulations have comparable pot lives and curing characteristics as measured by Shore A Hardness Test Method. The only difference occurs from the longer tack-free time obtained for DMBA cured IGS. However the addition of 30 minutes is not considered significant.

3 - Initial adhesion to aluminum

Adhesion to aluminum and glass is evaluated with a 90-degree peel test performed on special specimens made of one piece of glass on which a 2-inch wide thin strip of aluminum is bonded. Initial adhesion means after 24 hours at room temperature.

The level of the peel strength itself is not the more important factor : nowadays a strength of 6-7 N/mm is considered as satisfactory. What is essential is to obtain a cohesive failure inside the layer of sealant.

In Table 5, the values obtained for IGS using the DMBA based curing system and without the coumarone indene (MCI) are also provided for comparison.

TABLE 5 -- Initial adhesion

| Catalyst | Peel strength, N/mm | Type of failure |
|---|---|---|
| PME | 9.0 | Cohesive |
| DMBA | 10.4 | Cohesive |
| DMBA no MCI | 3.6 | Adhesive to aluminum |

4- permeability

The change has no effect on the MVTR and argon permeability which remain at the same good levels :

| | | |
|---|---|---|
| MVTR | (DIN 53122-20°C) : | 1.0 g/m2/day |
| Argon permeability | (DIN 52344-20°C) : | 0.10 l/m2/day/atm |

Long term ageing tests on insulating glass units are currently under evaluation according to French specification NF-P 78-451. The final results are not yet available.

**CONCLUSION**

In conclusion a catalyst altogether as good as the phenyl mercury ester for polyurethanes based on hydroxylated polybutadiene has still to be found. Nevertheless it has been shown that, under certain conditions and with few modifications, the dimethylbenzylamine allows an acceptable level for all the essential characteristics of insulating glass sealants.

In case of necessity this solution may become industrially helpful after scaling up.
The raw material cost is not significantly different and therefore the cost quality efficiency in favor of polyurethanes in this application is not altered.

## REFERENCES

[1] US Patent 4,063,002 (1977)

[2] US Patent 4,430,489 (1984)

[3] US Patent 4,716,070 (1987)

[4] US Patent 4,780,521 (1988)

# Studies Related to the Long-Term Performance and Aging of Sealants and Gaskets

J. L. Beasley[1] and M. R. Jenkins[1]

# THE EFFECT OF ARTIFICIAL WEATHERING AND MOVEMENT ACCOMMODATION TESTING ON BUILDING SEALANTS

**REFERENCE:** Beasley, J. L. and Jenkins, M. R., **"The Effect of Artificial Weathering and Movement Accommodation Testing on Building Sealants,"** Science and Technology of Building Seals, Sealants, Glazing, and Waterproofing: Fifth Volume, ASTM STP 1271, Michael A. Lacasse, Ed., American Society for Testing and Materials, 1996.

## ABSTRACT

This paper describes the effect of an ultra violet radiation, heat and condensation artificial ageing procedure on the ability of a range of building sealants to survive a movement accommodation test based on the ISO 9047 test method. Sealants stored under standard laboratory conditions were also subjected to the movement accommodation test.

It is shown that sealants which initially pass the movement accommodation test will continue to pass after up to 24 weeks of exposure to the artificial weathering, or storage in standard conditions. Sealants that initially do not survive the movement test can improve their performance after exposure to artificial weathering or storage in standard conditions.

Tensile tests were carried out on sealants that survived the movement test and on static control specimens. It has been found that the movement test, and to a lesser extent the artificial weathering, can reduce the width of the test samples. It has also been found that the movement test can decrease the force produced by most one part sealants when extended to a width of 15mm, nominally 25% strain.

No relationship between changes in sample width or force at 15mm width and the ability to survive the movement test has been found.

**KEYWORDS:** sealant, width, modulus, weathering, movement accommodation

[1] Building Research Establishment, Garston, Watford, WD2 7JR, United Kingdom.

## INTRODUCTION

Subjecting a sealant to a cyclic movement test after a period of weathering, either artificial or natural, may give an indication of continuing ability to perform the function of sealing a joint. The use of a weathering regime may change some of the mechanical properties of the sealant. The cyclic movement test used to assess the sealant's continuing ability to seal a joint may also affect the mechanical properties of the sealant.

The work described here investigated the effect of an ageing regime of heat and UV radiation alternated with moisture in the form of condensation, followed by a test of movement accommodation based around the procedures described in ISO 9047,[1]. The changes in modulus brought about in freshly cured sealant by this artificial weathering regime over an eight week period have been shown to correlate well with those that occur in the temperate climate of southern England over a period of 52 weeks,[2,3].

This work was carried out to determine the effects that cyclic movement and artificial weathering have on the the mechanical properties of building sealants, both singly and in combination. It has also been possible to investigate the effect that artificial weathering has on the ability of building sealants to accommodate cyclic movement.

All test pieces that survived the movement accommodation test were subjected to a tensile test. The width of the sample was measured immediately before the tensile test was carried out. The force produced by the test piece was measured at a width of 15mm as this corresponds with the width to which the sealant was extended in the movement accommodation test.

Changes in the width of the sealant test piece will alter the amount of extension and compression applied to the sealant. The ISO 9047 test method assumes that the test piece has a width of 12mm. A decrease in this width will have the effect of increasing the amount of extension imposed and decreasing the amount of compression. An increase in sample width will decrease the amount of extension applied and increase the amount of compression.

Changes in the force measured at 15mm width will affect the stresses within the sealant bead and at the bondline with the substrate. It would be possible for changes in these properties to combine in ways that reinforce or cancel out each other's effects. A decrease in width combined with a decrease in modulus may leave the stresses found at 15mm width unchanged. Other combinations of changes in width and modulus can be devised and their possible effects on stress levels envisaged.

## EXPERIMENTAL

Seven sealants were used in this test programme. Six of the sealants are commercially available in the UK and one is a model two part polysulphide formulation. They are coded as follows:

| | | |
|---|---|---|
| P13 | Black | One part polysulphide to BS 5215[4] |

| | | |
|---|---|---|
| 2P10 | Grey | Two part polysulphide to BS 4254[5] |
| 2P16 | Brown | Model two part polysulphide |
| | | |
| S6 | Grey | Silicone to BS 5889[6] |
| S15 | Grey | Silicone to BS 5889[6] |
| | | |
| U1 | Grey | One part polyurethane |
| U6 | Grey | One part polyurethane |

The test pieces used consisted of a 50mm x 12mm x 12mm bead of sealant held between one cement mortar and one aluminium substrate. Test samples of each sealant type were produced using sealant from the same batch and the substrates were primed according to the relevant manufacturer's instructions. Samples were cured according to ISO 9047 method B. The samples were stored for 28 days at 23±2°C and 50±5% rh and then subjected to the following cycle of heat and immersion that was repeated three times:

2 days in oven at 70±2°C,
1 day in deionised water at 23±1°C,
3 days in oven at 70±2°C,
1 day in deionised water at 23±1°C.

The modified ISO 9047 movement accommodation test was:

Test specimens extended and compressed at a rate of strain of 5mm/min.

First week:

| | |
|---|---|
| Day 1. | 3 hours cooling to -20±2°C and subsequent extension to 15mm width. 21 hours maintained extension at -20±2°C. |
| Day 2. | Release of extension, 3 hours heating to 70±2°C and subsequent compression to 9.75mm width, 21 hours maintained compression at 70±2°C. |
| Day 3. | Release of compression and procedure of day 1. |
| Day 4. | Procedure of day 2. |
| Days 5 to 7. | Release of compression, store at 23±2°C without application mechanical force. |

Second week: procedure of the first week.

After the curing procedure, the samples were randomly allocated to the following exposure conditions:

1. 15 test pieces stored at 23±2°C and 50±5% rh. Sets of five samples were removed at intervals of 8, 16 and 24 weeks and subjected to a tensile test.

2. 15 test pieces stored at 23±2°C and 50±5% rh. Sets of five samples were removed at intervals of 8, 16 and 24 weeks and were subjected to the modified ISO 9047 strain cycling procedure followed by a tensile test.

3. 15 test pieces exposed in an ultra violet radiation and condensation weathering apparatus, fitted with 313nm lamps, set to a cycle of 4 hours UVB radiation at 70°C followed by 4 hours of condensation at 40°C. Sets of five samples were removed at intervals of 8, 16 and 24 weeks and subjected to a tensile test.

4. 15 test pieces exposed in an ultra violet radiation and condensation weathering apparatus, fitted with 313nm lamps, set to a cycle of 4 hours UVB radiation at 70°C followed by 4 hours of condensation at 40°C. Sets of five samples were removed at intervals of 8,16 and 24 weeks and subjected to the modified ISO 9047 strain cycling procedure followed by a tensile test.

Five test pieces were subjected to the modified ISO 9047 test immediately after the ISO 9047 method B curing procedure.

Before the tensile test the widths of the test pieces were measured using digital calipers. The calipers were accurate to ±0.005mm. Samples were pulled to break on a Hounsfield H2000 tensile testing machine at a rate of extension of 5.0±0.1mm $min^{-1}$. The load was recorded at 0.1±0.01mm intervals of extension with an accuracy of ±1.0 newton. Because the cross sectional area of each sample was the same it was not felt necessary to convert the load values to stresses.

## RESULTS AND DISCUSSION

### Cyclic Movement

The regime used to impose the cyclic movement was based on, but different from that described in ISO 9047. An extension of 3mm and compression of 2.25mm was used. Based on the nominal width of 12mm this gives an extension of 25% and a compression of 18.75%. All other parts of the regime were as described in ISO 9047. To fully comply with ISO 9047 the amount of compression applied should be the same as that applied in tension. However because of the shape and size of the substrates used it was not possible to compress the test piece by 3mm. The substrate size did not allow enough clearance between the jaws of the sample holder and the sealant bead. At compressions of over 2.25%mm the jaws would have impinged on the sealant bead and possibly initiated a cohesive failure from the marks left on the sealant bead. The results from the movement accommodation tests for artificially weathered test pieces and those exposed to standard conditions are shown in Tables 1 and 2.

The two polyurethane sealants , U1 and U6, and both of the two part polysulphides seemed to be unaffected in their ability to survive the cyclic movement test whether or not they are exposed to the artificial weathering regime. However there was a suggestion that the unweathered U1 sealant was just beginning to lose its ability to accommodate the movement after 24 weeks under control conditions.

P13 showed an increase in the number of samples surviving the cyclic movement regime after eight weeks in the ultra violet radiation and condensation weathering apparatus. Subsequently, after 16 and 24 weeks in the apparatus, only one sample out of five survived. Of the P13 test pieces subjected to the movement regime after the initial curing period none survived intact. After eight weeks in control conditions two samples survived and after 16 weeks and 24 weeks three samples survived, an improvement.

The two silicone sealants showed broadly similar patterns in their ability to accommodate the movement regime after storage in both the ultra violet radiation and condensation weathering apparatus and in controlled conditions. Both sealants, whether stored in the ultra violet radiation and condensation weathering apparatus or in controlled conditions, were initially unable to perform well in the cyclic movement test. The improvement in survival was more marked in S6.

It would appear that if a sealant is able to survive an ISO 9047 type movement accommodation test after the conditioning procedure used here, then the weathering procedure used in this programme of work will have very little effect on the sealant's continuing ability to accommodate the movement regime for up to 24 weeks of exposure. Sealants that perform poorly initially, P13, S6 and S15, will have their performance modified by the artificial weathering procedure compared with storage under control conditions. However the change that occurs is not consistent across the sealant types.

**Table 1**

Number of weathered samples surviving after cyclic movement regime

| Week | P13 | 2P10 | 2P15 | S6 | S15 | U1 | U6 |
|---|---|---|---|---|---|---|---|
| 0 | NA | NA | NA | NA | NA | NA | NA |
| 8 | 3 | 5 | 5 | 1 | 0 | 5 | 5 |
| 16 | 1 | 5 | 5 | 2 | 0 | 5 | 5 |
| 24 | 1 | 5 | 5 | 4 | 3 | 5 | 5 |

NA = Not Applicable, weathered samples not available at start of experiment.

**Table 2**

Number of unweathered samples surviving after cyclic movement regime

| Week | P13 | 2P10 | 2P15 | S6 | S15 | U1 | U6 |
|---|---|---|---|---|---|---|---|
| 0 | 0 | 5 | 5 | 1 | 0 | 5 | 5 |
| 8 | 2 | 5 | 5 | 4 | 0 | 5 | 5 |
| 16 | 3 | 5 | 5 | 5 | 1 | 5 | 5 |
| 24 | 3 | 5 | 5 | 1 | 3 | 4 | 5 |

## WIDTH BEFORE TENSILE TEST

The nominal width of the samples tested is 12mm but as can be seen in Table 3 this width is rarely found in any of the sets of samples. In general it can be seen that the average widths of the test pieces is less than the nominal width.

The extension and compressions described in the ISO 9047 test method assume that the width of the sample is 12mm. A decrease in this width will lead to the sample being subject to rather more extension and rather less compression that envisaged.

To determine whether significant changes had occurred in the width of the test pieces after exposure to the different movement and exposure regimes a t-test was performed on the data from weeks 8, 16 and 24. The results are shown in Tables 4, 5, 6 and 7.

### Changes in width between static and strain cycled sealant.

Table 4 shows whether the difference in widths between static and strain cycled samples that were stored in the ultra violet radiation and condensation weathering apparatus are significant. Where no comparison is made this is because insufficient test pieces survived the strain cycling procedure. It can be seen that of the 16 comparisons made 12 significant differences, at the 95% or 99% level, were found. In every instance in Table 4 where a significant difference is seen, a decrease in the width of the strain cycled test piece is found.

The results from the same procedure carried out between static and strain cycled sealants that were stored under standard conditions are shown in Table 5. Of the 18 comparisons made 13 significant differences were found. In this instance all the significant differences are at the 99% level whereas in the previous comparisons the level of significance was divided equally between the 99% and 95% levels. Again all the differences in width showed that the movement accommodation test tended to decrease the width of the sample. The tendency to decrease the width of the sample appears to be rather more pronounced in the samples that were stored under controlled conditions than in the ultra violet radiation and condensation weathering apparatus.

It is possible that the periods of compression at 70°C that occur in the ISO 9047 test will produce a degree of compression set in a sealant and that a less well cured sealant would be more susceptible to this happening. The tendency towards differences in width being observed at lower levels of significance for sealants that were stored in the ultra violet radiationand condensation weathering apparatus is consistent with them being at a more advanced state of cure, and therefore less susceptible to acquiring a compression set during the period of compression in the ISO 9047 test.

### Effect of storage condition on test piece width

It is also possible to investigate whether there are any significant differences in width caused by exposure conditions. Table 6 shows comparisons in sample width for static samples. In this instance 21 comparisons were made and eight differences significant at the 99% level and one at the 95% level were found. Most of the significant changes in width occurred in the polysulphide sealants. All of the changes observed showed that storage in the ultra violet radiation and condensation weathering apparatus tended to cause a decrease in the width of the test pieces.

The final comparison of widths of test pieces was carried out between samples that had been subjected to the movement accommodation test, and those that had been stored in the ultra violet and condensation weathering apparatus or in standard conditions, Table 7. Fifteen comparisons were made and only two were found to be significant at the 95% level. Unlike the previous three comparisons no consistent pattern in the differences in width were found. This suggests that although storage in an ultra violet radiation and condensation weathering apparatus can have a tendency to reduce the width of the static test piece as shown in Table 6 the differences produced by the artificial weathering will be obscured by subjecting a sealant to the ISO 9047 movement accommodation test.

**Table 3** Average width of test pieces prior to tensile testing / mm.

| Conditions | Weeks | P13 | 2P10 | 2P16 | S6 | S15 | U1 | U6 |
|---|---|---|---|---|---|---|---|---|
| UV(B) radiation (70°C) and condensation (40°C) followed by movement accommodation test procedure. | 8 | 11.36 (0.08) | 11.55 (0.12) | 11.40 (0.25) | --- | --- | 11.69 (0.03) | 11.73 (0.12) |
| | 16 | --- | 11.55 (0.13) | 11.16 (0.19) | 11.62 (0.02) | --- | 11.75 (0.08) | 11.76 (0.12) |
| | 24 | --- | 11.40 (0.07) | 11.34 (0.12) | 11.56 (0.07) | 11.53 (0.09) | 11.69 (0.09) | 11.78 (0.09) |
| Stored at 23°C and 50% rh followed by movement accommodation test procedure. | 8 | 11.45 (0.18) | 11.67 (0.21) | 11.27 (0.25) | 11.54 (0.14) | --- | 11.65 (0.04) | 11.72 (0.10) |
| | 16 | 11.27 (0.09) | 11.58 (0.04) | 11.21 (0.08) | 11.54 (0.06) | --- | 11.71 (0.15) | 11.79 (0.10) |
| | 24 | 11.30 (0.12) | 11.56 (0.12) | 11.51 (0.12) | --- | 11.81 (0.08) | 11.69 (0.03) | 11.78 (0.16) |
| UV(B) radiation (70°C) and condensation (40°C) only. | 8 | 11.74 (0.13) | 11.87 (0.07) | 11.72 (0.06) | 11.84 (0.07) | 11.58 (0.07) | 11.83 (0.11) | 12.01 (0.06) |
| | 16 | 11.41 (0.10) | 11.76 (0.10) | 11.27 (0.08) | 11.82 (0.09) | 11.73 (0.10) | 11.76 (0.03) | 11.86 (0.08) |
| | 24 | 11.17 (0.25) | 11.76 (0.03) | 11.49 (0.07) | 11.70 (0.03) | 11.68 (0.09) | 11.82 (0.06) | 11.93 (0.04) |
| Stored at 23°C and 50% rh only. | 8 | 12.00 (0.10) | 11.91 (0.03) | 11.88 (0.04) | 11.84 (0.06) | 11.85 (0.07) | 11.78 (0.04) | 12.00 (0.09) |
| | 16 | 11.91 (0.07) | 11.84 (0.10) | 11.90 (0.10) | 11.79 (0.07) | 11.75 (0.06) | 11.71 (0.08) | 11.92 (0.10) |
| | 24 | 11.99 (0.08) | 11.93 (0.05) | 11.95 (0.03) | 11.73 (0.03) | 11.79 (0.05) | 11.81 (0.06) | 11.96 (0.03) |

Key: --- One sample or no samples survived the movement accommodation test.
Value inside brackets () is the sample standard deviation.

**Table 4**

Comparisons of width before tensile test of static and strain cycled test pieces that were stored in a Q-UV weathering apparatus

| Week | P13 | 2P10 | 2P16 | S6 | S15 | U1 | U6 |
|---|---|---|---|---|---|---|---|
| 8 | ** ↓ | ** ↓ | * ↓ | - | - | * ↓ | ** ↓ |
| 16 | - | * ↓ | N | * ↓ | - | N | N |
| 24 | - | ** ↓ | * ↓ | ** ↓ | N | * ↓ | ** ↓ |

Key: ** Significant at 99% level
* Significant at 95% level
↓ Width tends to decrease with respect to static samples
N No significant change in width with respect to static samples
- Only one sample or no samples survived the movement accommodation test.

**Table 5**

Comparisons of width before tensile test of static and strain cycled test pieces that were stored at 23°C and 50% rh.

| Week | P13 | 2P10 | 2P16 | S6 | S15 | U1 | U6 |
|---|---|---|---|---|---|---|---|
| 8 | ** ↓ | N | ** ↓ | ** ↓ | - | ** ↓ | ** ↓ |
| 16 | ** ↓ | ** ↓ | ** ↓ | ** ↓ | - | N | N |
| 24 | ** ↓ | ** ↓ | ** ↓ | - | N | ** ↓ | N |

Key: ** Significant at 99% level
↓ Width tends to decrease with respect to static samples
N No change in width with respect to static samples
- Only one sample or no samples survived the movement accommodation test.

**Table 6**

Comparisons of width before tensile test of static test pieces that were stored in a Q-UV weathering apparatus and those stored at 23°C and 50% rh.

| Week | P13 | 2P10 | 2P16 | S6 | S15 | U1 | U6 |
|---|---|---|---|---|---|---|---|
| 8 | ** ↓ | N | ** ↓ | N | ** ↓ | N | N |
| 16 | ** ↓ | N | ** ↓ | N | N | N | N |
| 24 | ** ↓ | ** ↓ | ** ↓ | N | * ↓ | N | N |

Key: ** Significant at 99% level
* Significant at 95% level
↓ Width decreases with respect to control storage samples
N No change in width with respect to control storage samples

**Table 7**

Comparisons of width before tensile test of strain cycled test pieces that were stored in a Q-UV weathering apparatus and those stored at 23°C and 50% rh.

| Week | P13 | 2P10 | 2P16 | S6 | S15 | U1 | U6 |
|---|---|---|---|---|---|---|---|
| 8 | N | N | N | - | - | N | N |
| 16 | - | N | N | N | - | N | N |
| 24 | - | * ↓ | N | - | * ↓ | N | N |

Key: * Significant at 95% level
↓ Width decreases with respect to control storage samples
N No significant change in width with respect to control storage samples
- Only one sample or no samples survived the movement accommodation test.

**FORCE AT 15mm WIDTH**

The forces measured at 15mm width after the different lengths of time in the different exposure conditions are shown in Table 8.

**Changes in force at 15mm width between static and strain cycled sealant.**

Table 9 shows whether the differences in force at 15mm width between static and strain cycled samples that were stored in the ultra violet radiation and condensation weathering apparatus are significant. Where no comparison is made this is because insufficient samples survived the strain cycling procedure. It can be seen that 16 comparisons were made of which seven were significant at the 95% or 99% level.

All but one of the significant changes, the exception to this trend being 2P10 at 24 weeks, show decreases in the force produced by samples that had been subjected to the ISO9047 movement accommodation test.

Table 10 shows the same comparison as above but for sealants that were stored at 23°C and 50% rh. Eighteen comparisons were made. Twelve significant differences were found at the 99% or 95% level. All the one part sealants showed decreases in the force at 15mm width after the movement accommodation test when compared with static test pieces whereas both the two part polysulphides showed increases in force.

It appears that the movement accommodation test tends to reduce the modulus of the one part sealants and increase that of the two part polysulphides, although not all the changes are significant. The levels of significance that are found indicate that this tendency is more pronounced in the test pieces that were stored under standard conditions.

**Table 8**
Mean force at 15mm width /newtons

| Conditions | Weeks | P13 | 2P10 | 2P16 | S6 | S15 | U1 | U6 |
|---|---|---|---|---|---|---|---|---|
| UV(B) radiation (70°C) and condensation (40°C) followed by movement accommodation test procedure. | 8 | 37.3 (9.2) | 146.3 (11.4) | 41.8 (4.3) | --- | --- | 99.0 (4.9) | 77.4 (8.1) |
| | 16 | --- | 152.5 (8.0) | 49.1 (2.3) | 50.2 (0.3) | --- | 98.0 (13.9) | 87.4 (13.9) |
| | 24 | --- | 163.1 (4.9) | 41.8 (1.1) | 51.7 (3.2) | 52.4 (1.6) | 106.2 (5.0) | 80.6 (8.7) |
| Stored at 23°C and 50% rh followed by movement accommodation test procedure. | 8 | 30.6 (1.4) | 146.2 (16.4) | 49.9 (5.3) | 44.5 (3.1) | --- | 118.8 (13.5) | 72.1 (5.3) |
| | 16 | 34.3 (7.7) | 150.8 (7.4) | 57.4 (2.4) | 48.3 (1.3) | --- | 123.3 (12.3) | 78.8 (6.7) |
| | 24 | 41.9 (2.4) | 153.9 (5.0) | 54.7 (2.7) | --- | 46.1 (4.7) | 133.4 (10.7) | 87.4 (6.5) |
| UV(B) radiation (70°C) and condensation (40°C) only. | 8 | 52.0 (5.9) | 135.0 (2.7) | 45.3 (1.8) | 53.0 (7.2) | 58.8 (3.7) | 127.5 (13.6) | 84.1 (5.7) |
| | 16 | 57.0 (2.3) | 146.2 (4.1) | 49.8 (2.2) | 65.2 (8.2) | 66.2 (4.5) | 130.6 (12.6) | 88.2 (2.4) |
| | 24 | 70.6 (6.4) | 147.9 (4.2) | 41.4 (0.7) | 60.8 (9.6) | 67.4 (6.3) | 119.4 (10.2) | 94.6 (8.7) |
| Stored at 23°C and 50% rh only. | 8 | 47.2 (12.9) | 129.0 (6.6) | 42.1 (0.7) | 57.3 (5.3) | 56.6 (2.9) | 140.7 (6.6) | 83.9 (5.9) |
| | 16 | 47.0 (10.7) | 132.4 (6.2) | 47.6 (1.1) | 57.6 (6.0) | 66.1 (4.2) | 143.8 (12.0) | 87.8 (13.9) |
| | 24 | 52.3 (18.8) | 141.0 (4.5) | 48.4 (1.6) | 58.4 (5.4) | 62.8 (6.2) | 161.7 (6.4) | 87.5 (6.6) |

Key : --- One sample or no samples survived the movement accommodation test.
Value inside brackets () is the sample standard deviation

**Table 9**

Comparisons of force at 15mm width of static and strain cycled test pieces that were stored in a Q-UV weathering apparatus.

| Week | P13 | 2P10 | 2P16 | S6 | S15 | U1 | U6 |
|---|---|---|---|---|---|---|---|
| 8 | * ↓ | N | N | - | - | ** ↓ | N |
| 16 | - | N | N | N | - | ** ↓ | N |
| 24 | - | ** ↑ | N | N | ** ↓ | * ↓ | ** ↓ |

**Table 10**

Comparisons of force at 15mm width of static and strain cycled test pieces that were stored at 23°C and 50% rh.

| Week | P13 | 2P10 | 2P16 | S6 | S15 | U1 | U6 |
|---|---|---|---|---|---|---|---|
| 8 | N | N | * ↑ | ** ↓ | - | * ↓ | * |
| 16 | N | ** ↑ | ** ↑ | * ↓ | - | * ↓ | N |
| 24 | N | ** ↑ | ** ↑ | - | ** ↓ | ** ↓ | N |

**Table 11**

Comparisons of force at 15mm width of static test pieces that were stored in a Q-UV weathering apparatus and those stored at 23°C and 50%rh.

| Week | P13 | 2P10 | 2P16 | S6 | S15 | U1 | U6 |
|---|---|---|---|---|---|---|---|
| 8 | N | N | ** ↑ | N | N | N | N |
| 16 | N | ** ↑ | N | N | N | N | N |
| 24 | N | * ↑ | ** ↓ | N | N | ** ↓ | N |

Key:
- ** Significant at 99% level
- * Significant at 95% level
- ↓ Force decreases with respect to control storage samples
- ↑ Force increases with respect to control storage samples
- N No significant change in force with respect to control storage samples

**Table 12**

Comparisons of force at 15mm width of strain cycled test pieces that were stored in a Q-UV weathering apparatus and those stored at 23°C and 50% rh.

| Week | P13 | 2P10 | 2P16 | S6 | S15 | U1 | U6 |
|---|---|---|---|---|---|---|---|
| 8 | N | N | * ↓ | - | - | * ↓ | N |
| 16 | - | N | ** ↓ | N | - | * ↓ | N |
| 24 | - | * ↑ | ** ↓ | - | N | ** ↓ | N |

Key:
- ** Significant at 99% level
- * Significant at 95% level
- ↓ Force decreases with respect to control storage samples
- ↑ Force increases with respect to control storage samples
- N No significant change in force with respect to control storage samples
- \- Only one sample or no samples survived the movement accommodation test.

**Effect of storage condition on force at 15mm width.**

The differences in force at 15mm width between static samples stored under standard conditions and those exposed in the ultra violet radiation and condensation weathering apparatus are shown in Table 11. Twenty one comparisons were made and four significant differences found at the 99% level and one at the 95% level. There does not appear to be a general trend in the direction of the change in force across all the sealants tested.

The differences in force at 15mm width between samples subjected to the ISO 9047 movement accommodation test that were stored under standard conditions and those exposed in the ultra violet radiation and condensation weathering apparatus are shown in Table 12. Fifteen comparisons were made and three significant differences at the 99% level and four significant differences at the 95% level were found. Where a significant difference occurs the samples stored in the ultra violet radiation and condensation apparatus tend to show a decrease in force at 15mm.

**Width, Force and Movement Accommodation Test Survivability.**

Attempts have been made using the data presented here to find a pattern that would relate changes in survival in the ISO 9047 test to changes in either sample width or force at 15 mm width. No such relationship has been discovered.

**CONCLUSIONS**

1. The use of the UV-B/Condensation weathering regime described here appears not to affect the ability of a sealant to survive an ISO 9047 type movement test where the sealant has been able to survive the test immediately after curing to ISO 9047 method B.

2. The ISO 9047 strain cycling procedure tends to reduce the width of the sealant test specimen when compared with test specimens that have not been subjected to the procedure. This change in width will occur for samples stored under control conditions and for those subjected to the ultra violet radiation and condensation weathering procedure used in this work.

3. The ultraviolet radiation and condensation weathering regime can also produce a decrease in the width of a sealant test piece that is not subjected to the ISO 9047 movement accommodation test.

4. The ISO 9047 movement accommodation test tends to decrease the modulus of the one part sealants whether they have been stored previously under standard laboratory conditions or exposed in a Q-UV weathering apparatus.

5. Changes in the properties of building sealant can be associated with artificial weathering and cyclic movement. Attempts to produce a test method for assessing durability will have to take account of both of these influences.

6. The ISO 9047 test is established as test for assessing the movement accommodation capabilities of a building sealant that has undergone an initial curing procedure. The test may have to be modified if it is to be used in conjunction with artificial ageing procedures.

## REFERENCES

1. International Standard Specification ISO9047, 1989, Building construction - Sealants - Determination of adhesion/cohesion properties at variable temperatures, International Organisation for Standardisation, Case postale 56, CH-1211 Geneve 20, Switzerland.

2. Beech, J.C. and Beasley, J.L., "Evaluation of Cure and Durability Aspects of Building Sealants," Science and Technology of Building Seals, Sealants, Glazing and Waterproofing: Second Volume, ASTM STP 1200, Jerome M. Klosowski, Ed. American Society for Testing and Materials, Philadelphia, 1992.

3. Beech, J.C. and Beasley, J.L., "Effects of natural and artificial weathering on building sealants", Science and Technology of Building Seals, Sealants, Glazing and Waterproofing: Fourth Volume, ASTM STP 1243, D. Nicastro, Ed. American Society for Testing and Materials, Philadelphia, 1994.

4. British Standard Specification BS5215, 1983, One Part Gun Grade Polysulphide Based Sealants, BSI, London, 1983.

5. British Standard Specification BS4254, 1986, Two Part Polysulphide Based Sealants, BSI, London, 1986.

6. British Standard Specification BS5889, 1989, One Part Gun Grade Silicone Based Sealants, BSI, London, 1989.

Michael A. Lacasse[1], Jennifer E. Bryce[1], and James C. Margeson[1]

# EVALUATION OF CYCLIC FATIGUE AS A MEANS OF ASSESSING THE PERFORMANCE OF CONSTRUCTION JOINT SEALANTS: Polyurethane Sealants

---

**REFERENCE:** Lacasse[1], M.A., Bryce[1], J.E., and Margeson[1], J.C., **"Evaluation of Cyclic Fatigue as a Means of Assessing the Performance of Construction Joint Sealants: Polyurethane Sealants"**, Science and Technology of Building Seals, Sealants, Glazing, and Waterproofing: Fifth Volume, ASTM STP 1271, Michael A. Lacasse, Ed., American Society for Testing and Materials, 1996.

**ABSTRACT:** Polyurethane based high performance construction sealants were subjected to cyclical movements using manually adjusted vices until rupture of the specimens occurred. Sealant fatigue behavior was compared with the fatigue behavior of elastomers and rubbers in order to gain a more quantitative understanding of sealants subjected to fatigue testing. Furthermore, failure of these sealant specimens during cyclic fatigue testing allowed for a comparison with previously reported results obtained from tests on both silicone and polysulfide sealants. The methods described in this paper provide a basis from which to assess the long-term performance of different types of sealants.

**KEY WORDS:** construction sealants, cyclic fatigue, cyclical movement, polyurethane sealant, polysulfide sealant, sealant failure, sealant fatigue, sealant long-term performance, silicone sealant.

---

1. Research Officer, Laboratory Assistant and Technical Officer respectively, National Research Council Canada, Institute for Research in Construction, Materials Laboratory, Ottawa, Ontario, Canada, K1A 0R6;

## INTRODUCTION

The rubber-like flexibility of building joint sealants enables them to respond to and withstand repeated cycles of extension and compression as imposed on them by adjacent building elements while simultaneously sealing the joint; they prevent the passage of water, dust, and air through the building envelope. Hence the choice of a sealant product is based primarily on its ability to accommodate movement. To what extent can a product expect to accommodate movement and perform without rupture when subjected to repeated cycles of joint movement at or near its movement capability, and, what test methods should be employed to assess their degree of performance? These issues have been the subject of a number of investigations in recent years primarily because the premature failure of products can be a costly endeavor since the cost of sealant replacement and the repair of damage may be considerably more than that of the initial installation costs.

Previous investigations by Karpati [1-6] have shown that cyclic joint movement is the most important service factor influencing sealant performance. Correlation has been demonstrated between the performance of a two-part polysulfide sealant cycled on a strain-cycling outdoor exposure rack [7] and subjected to laboratory fatigue testing using manually-adjusted vises [4,8]. In addition, the laboratory method allowed the actual time to failure to be accelerated, and the number of cycles to each mode of failure was consistent with those of sealant specimens exposed outdoors. Work undertaken by Lacasse [9] in the lab has shown that fatigue behavior of polyurethane products can likewise be extrapolated to provide an indication of the performance potential of these sealants. In this work [9], an analysis of the results allowed for parallels to be drawn between the fatigue behavior of sealants and that of other elastomers according to models of fatigue in rubbers put forward by Bartenev [10]. It was shown that sealant fatigue behavior can be characterized using the same principles that models fatigue in elastomeric compounds according to the following relation [9,10]:

$$N\varepsilon^{m} = C \qquad (1)$$

In which, $\varepsilon$ is the predetermined maximum deformation of the sample (i.e. the amplitude of movement of the sealant specimen), N is the number of cycles to failure, and both m and C are constants related to the intrinsic fatigue properties of the material. Of these two constants neither are dependent on N, and m is neither dependent on the temperature nor the method of testing.

If results for sealant fatigue are plotted according to the relation provided by Bartenev, in which log N (cycles to failure) is shown as a function of the log strain amplitude ($\varepsilon$), a linear relationship occurs, as shown in Figure 1 below. A linear regression analysis of the data provides a means to determine the constants m and C that can then be used to predict cycles to failure at strain amplitudes found in service conditions.

$$\log N = \log C - m \log \varepsilon \qquad (2)$$

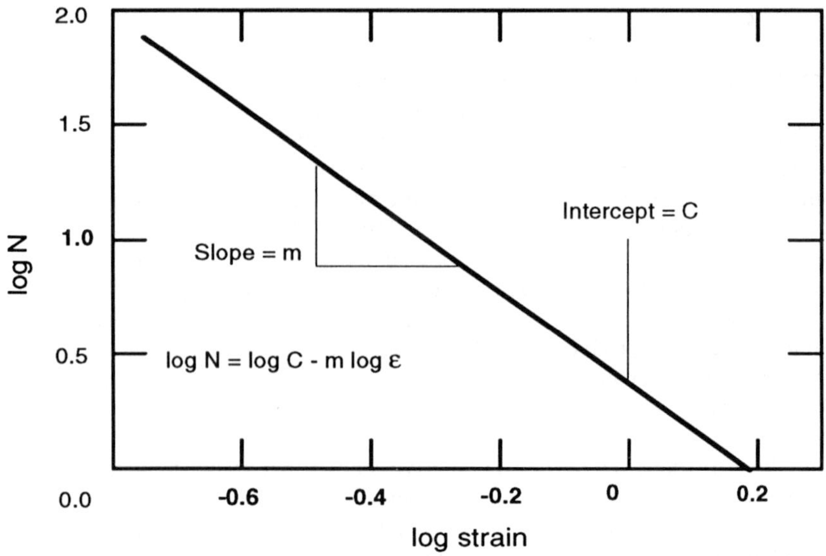


**Figure 1.** Log N (number of cycles to failure) as a function of log strain (log ε); m and C are constants related to the Bartenev equation, given in equation (2) above.

Based on this work [9,10] and that of Karpati [4], results were presented of a study in which the fatigue life of a series of five high performance silicone sealants was evaluated [11] using the laboratory method devised by Karpati [4]. From this study the failure data from different silicone based sealants produced similar results as those of Bartenev when a plot of log cycles to failure versus log strain was established as shown in Figure 1. Hence there exists a quantitative method of comparing the fatigue behavior of different sealants as well as for estimating the lifetime of sealed joints.

However, although these results demonstrated the potential usefulness of this method of analysis in the study of sealant fatigue, it was noted that precautions should be taken to ensure useful results, including [11]:

- initiating cyclic movement only on fully cured samples;
- limiting damage that occurs to the sealant both during casting and while being compressed;
- thorough testing using as large a number of samples as is practically possible;
- testing until failure, thus ensuring that it is not simply premature failures that are included in the results.

This study is a continuation of that undertaken on silicone based products and it focuses on comparing the fatigue life of three polyurethane based sealants using the same technique to induce fatigue to failure in model sealant beads as was used in the previous work [11]. Results from laboratory studies on a single component polyurethane sealant are compared to data obtained by Lacasse [9]. The performance potential of these sealants was determined using log N - log ε plots in an attempt to establish a relationship between sealant fatigue behavior and the theoretical model of Bartenev discussed in [9-11].

## EXPERIMENTAL

Materials:
Three commercial brand polyurethanes based products were evaluated in this study: a one-part moisture cured, gun-grade, non-sagging sealant (designated U02) having a gray color and two, two-component self-leveling products, also of gray color (designated L1 and L2 respectively). The later two products had been evaluated in a previous study (Lacasse [9]) and the results are presented here in order to offer a comparison of results obtained in the present study on product U02. Product L1 had a designated movement capability of ± 25% and L2 a capability of ± 50% according to information provided in the manufacturer's literature. Thirty-six specimens of products L1 and L2 respectively were evaluated in the previous study [9], whereas 104 specimens of U02 were tested in the present work (see Table 1 for details).

Procedure
Specimens were prepared using aluminum substrates having dimensions of 12.7 x 25.4 x 76.2 mm which were first mechanically cleaned and then solvent wiped with xylene. All products were applied to primed substrates in accordance with manufacturer's instructions. The product specimens were cast on silicone release paper and spacers were used to maintain the size of all sealant beads at 12.7 x 12.7 x 50.8 mm. In consideration of the work carried out in the previous study [9], and to ensure complete cure of the sealant products prior to the initiation of cyclic movement, specimens of product U01 were cured for approximately 48 days at 21 ± 2 °C and 50 ± 4% R.H. The specimens were then attached to hand-operated vises as previously described in [4,8].

The sealants tested by Lacasse [9] had been likewise cured for more than 48 days in the same conditions and set onto vises capable of accommodating up to six product specimens (Figure 2). Figure 3 shows the rack onto which a series of 18 of these vises were mounted. Two of these racks were used such that up to 216 specimens could be evaluated in cyclic movement at any one time. In both studies, sealant specimens were subjected to repeated cycles of extension and compression at a rate of 7.5 cycles per week [4].

As in the previous study on silicone sealants [11], the experiment on product U02 was broken into three phases based upon ranges of amplitude deformations, as shown in Table 1. The lower amplitudes of movement, ± 10% to ± 45%, were initiated in a first phase and to cover slightly higher amplitudes of movement a second phase included amplitudes ranging from ± 20% to ± 60%. This second phase also provided some overlap

**Figure 2.** Vise capable of accommodating up to six (6) model sealant specimens.

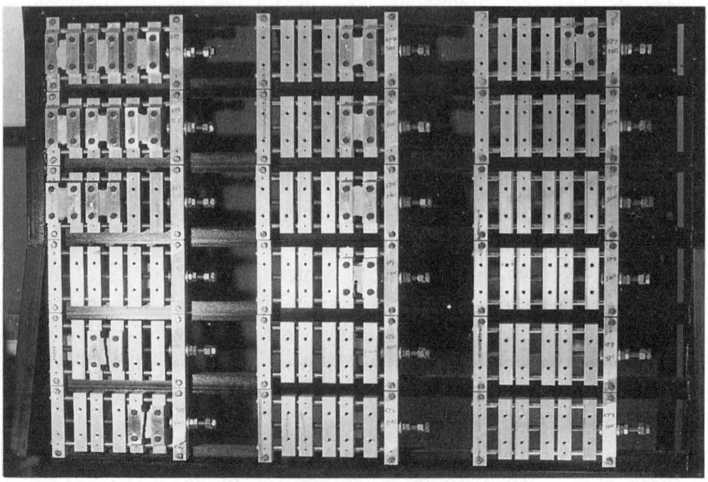

**Figure 3.** Rack onto which are mounted eighteen sealant test vises as shown in Figure 2.

in tests with specimens evaluated in phase one. In order to increase the reliability of the results a third phase was initiated and the products were again cycled from ± 20% to ± 60. For all amplitudes of movement of joint width in each phase, four specimens of each of the sealants were evaluated. Cyclic movements of extension and compression were imposed on the samples by manually adjusting the vises to give amplitudes of movement ranging from ± 10% to ± 60% of original joint width (12.7 mm), with amplitude increasing in 5% increments. All samples were started in compression (denoted by negative percentage) so that in extension (denoted by positive percentage) all samples had been through an integer number of cycles.

**Table 1. Number of specimens cycled at each amplitude**

| Amplitude of Movement (%) | U02 | | | L01 | L02 |
|---|---|---|---|---|---|
| | Phase I | Phase II | Phase III | | |
| 10 | 4 | | | | |
| 15 | 4 | | | | |
| 20 | 4 | 4 | 4 | | |
| 25 | 4 | 4 | 4 | 3 | 3 |
| 30 | 4 | 4 | 4 | 3 | 3 |
| 35 | 4 | 4 | 4 | 3 | 3 |
| 40 | 4 | 4 | 4 | 3 | 3 |
| 45 | 4 | 4 | 4 | 3 | 3 |
| 50 | | 4 | 4 | 3 | 3 |
| 55 | | 4 | 4 | 3 | 3 |
| 60 | | 4 | 4 | 3 | 3 |
| 65 | | | | 3 | 3 |
| 70 | | | | 3 | 3 |
| 75 | | | | 3 | 3 |
| 80 | | | | 3 | 3 |

For products L1 and L2, amplitudes of movement, imposed on the specimens by manually adjusting the vises, ranged from ± 25% to ± 80% of original joint width (12.7 mm), with amplitude increasing in 5% increments. For all amplitudes of movement of joint width three specimens of each of the sealant products were evaluated.

After each cycle, samples were examined for failure while still in extension, just before they were compressed. Photographs of the samples were taken while the sealant beads were in an extended state and these photographs were used as a visual record of the progression of failure. Sealant failure was visually identified as it progressed through essentially three stages of failure.

- Mode A failure: initial cracking/peeling at corners of sealant bead
- Mode B failure: cracking/peeling extended to nearly half the width of the specimen
- Mode C failure: complete rupture of sealant bead

Examples of these failures can be seen in Figures 4, 6, and 7. The number of cycles taken to reach each mode of failure was recorded and samples were removed from the testing after Mode C failure occurred. Results are reported as the average time to rupture (i.e., the number of cycles to Mode C failure) for a given amplitude of movement.

## RESULTS and DISCUSSION

Figure 4 is a photograph which shows a typical failure of a polyurethane (U02) specimen. The number of cycles to mode C failure (N) was used to calculate log N and the amplitude of movement of the specimens was used to determine the strain (ε) on the samples. The values of log N at each strain for each sealant specimen were averaged and log N versus log ε was then plotted for specimen U02 (Figure 5).

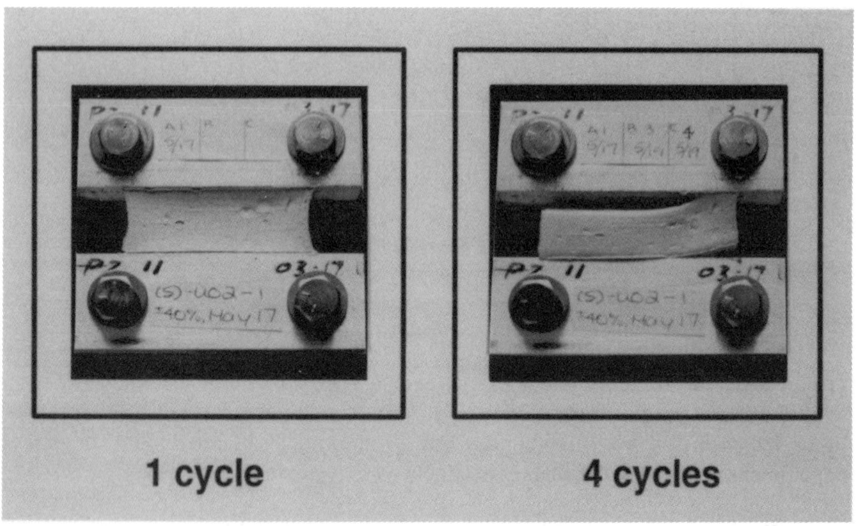


**Figure 4.** Mode of failure of product U02 subjected to a ± 40% movement amplitude after 1 and 4 cycles of movement respectively.

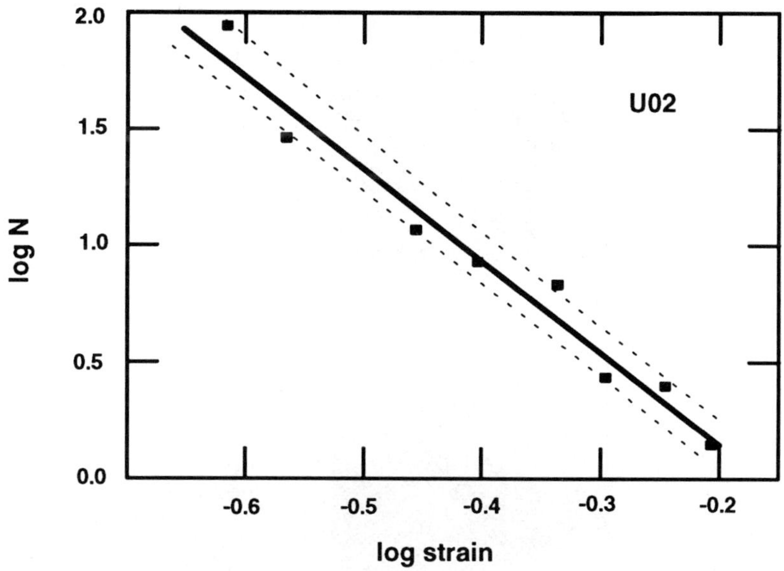


**Figure 5.** Log N (number of cycles to Mode C failure) as a function of log strain (± % amplitude of movement) for product U02.

Representative results for products L1 and L2 are depicted in Figure 6 and 7 respectively. In Figure 6, the progressive failure for a L1 specimen cycled at ± 70% amplitude movement from 2 to 4 cycles in shown, the last cycle indicative of Mode C failure. Figure 7 depicts similar progressive failure for product L2 cycled at the same ± 70% amplitude movement occurring over 7 cycles. The progressive nature of failure is evident from the rupture occurring at the interface between the sealant and the substrate after 2 cycles and thereafter after 3, 5 and 7 cycles of movement, again the last cycle indicative of Mode C failure. Corresponding plots of the number of cycles to failure as a function of the amplitude of strain are provided in Figures 8 and 9 for products L1 and L2 respectively. Both these plots indicate that the log of N is a linear function of the log of the strain amplitude, $\varepsilon$. In fact, for all polyurethane sealants evaluated in this study, regression of log N - log $\varepsilon$ plots (Figures 5,8,9) indicate a linear relationship as previously described in [11] for silicone sealants.

However, the strength of this relationship, based on 95% confidence limits, varied from sealant to sealant and, not unexpectedly, appeared to be strongest for sealants for which the largest number of samples were tested for the greatest number of cycles, as in the case of U02 (Figure 5). Polyurethane L1 (Figure 8) also shows a weaker log N - log $\varepsilon$ relationship than L2 (Figure 9), probably due to the limited number of samples tested.

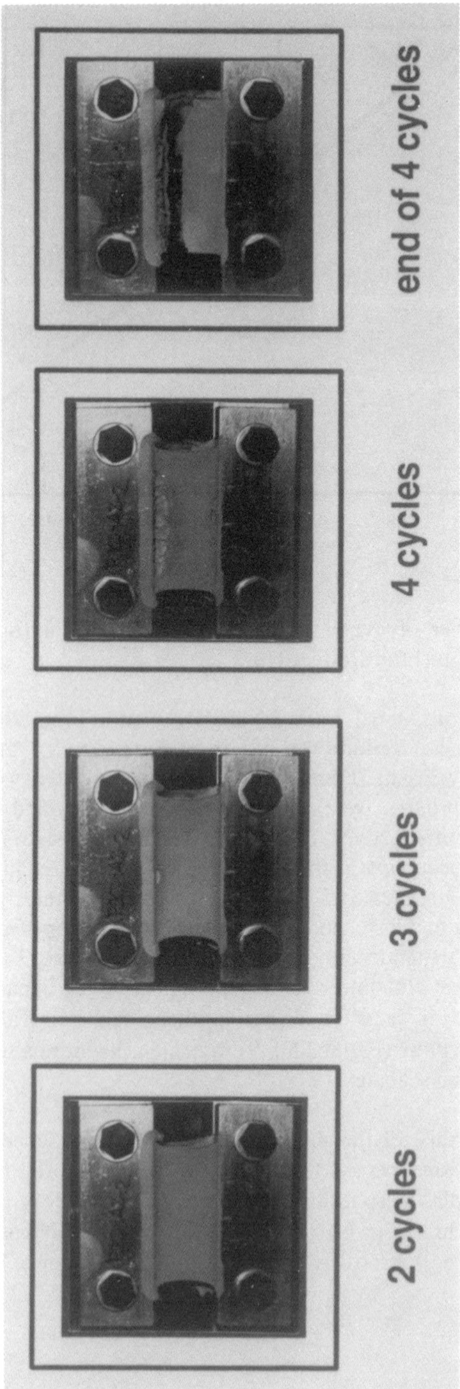


**Figure 6.** Progressive failure for a model specimen of product L1 cycled at ±70% amplitude movement showing rupture at the upper interface of the bead and substrate at 2 to 4 cycles, the last cycle being indicative of Mode C failure (complete bead rupture).

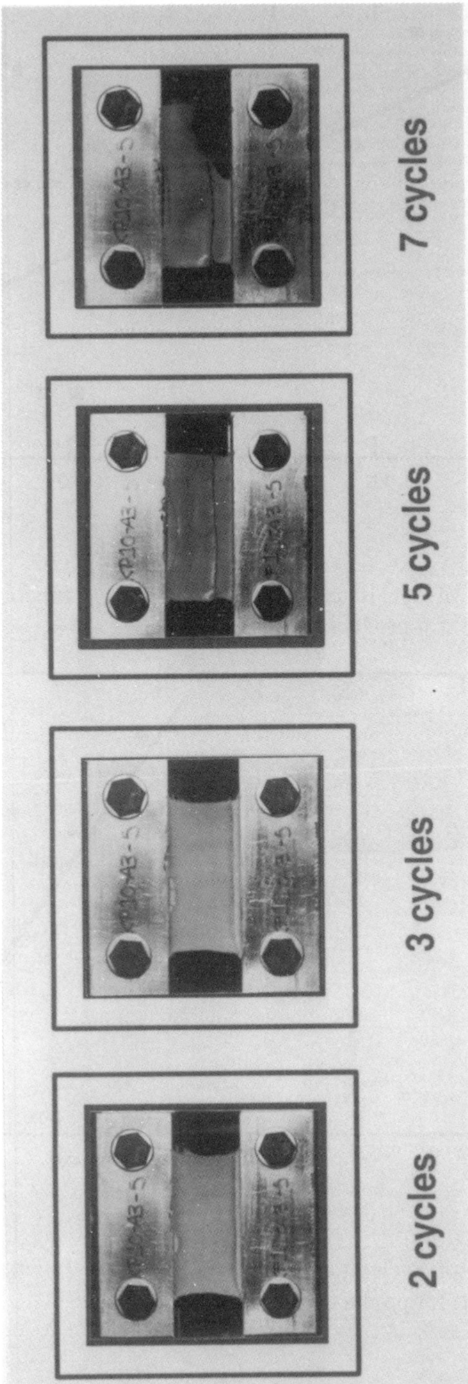


**Figure 7.** Similar progressive failure is shown for product L2 cycled at ±70% amplitude movement occurring over 7 cycles. The progressive nature of failure is evident from the rupture occurring at the interface between the sealant and the substrate after 2 cycles and thereafter after 3, 5 and 7 cycles of movement; the last cycle indicative of Mode C failure.

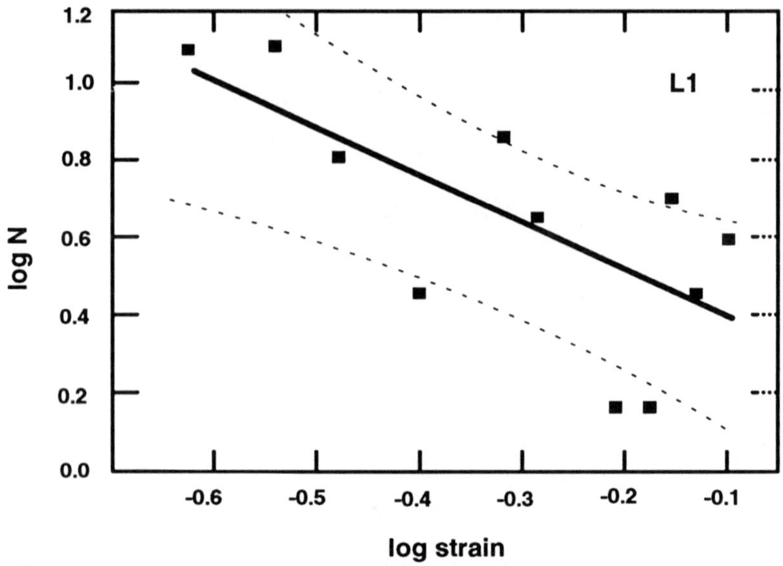


**Figure 8.** Log N (number of cycles to Mode C failure) as a function of log strain (± % amplitude of movement) for product L1.

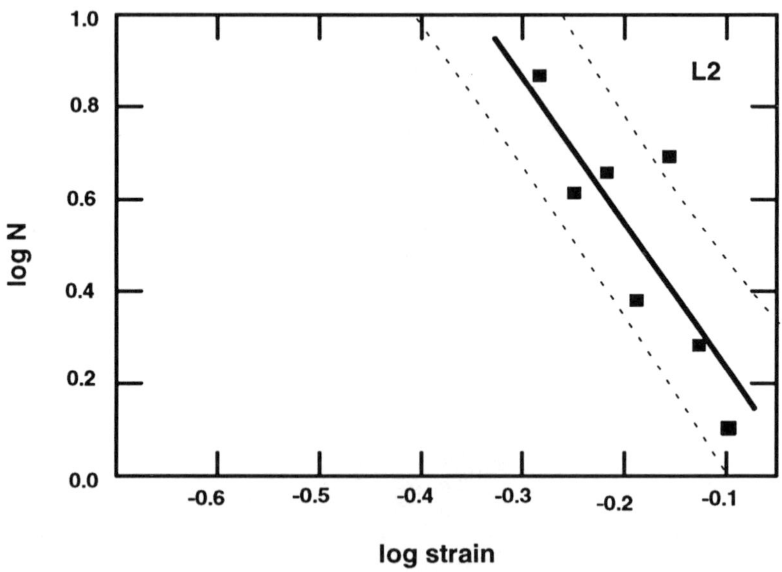


**Figure 9.** Log N (number of cycles to Mode C failure) as a function of log strain (± % amplitude of movement) for product L2.

Hence, from the plots provided in Figures 5, 8 and 9, it is evident that a linear relationship exists between log N and log ε, in which m and C are constants independent of both, ε, the strain amplitude, and N, the number of cycles to failure. The values for m and C can therefore be calculated from a linear regression of the log-log plot, and these are presented in Table 2 below for products L1, L2 and U02.

**Table 2. Value of constants from equations (1,2) as determined by linear regression**

| Sealant | m | C |
|---|---|---|
| **L1** | 1.60 | 1.16 |
| **L2** | 3.21 | 0.795 |
| **U02** | 4.78 | 0.101 |

An estimate of sealant lifetimes is made on the basis of the number of cycles to failure at a given amplitude, the average daily rate of movement of a 12.7 mm wide joint, and, the distance traveled in one cycle. This relationship is provided in equation 3 below.

$$L(_{lifetime}) = \frac{N \bullet d}{r} \qquad (3)$$

In which, N is the number of cycles to failure, d is the distance, in mm, traveled in one cycle, r, the average joint movement in mm / min., and L, the sealant product lifetime in minutes. The distance traveled, d, is simply calculated from the amplitude of movement through which the joint travels when undergoing one cycle of movement. The distance traveled for amplitudes of movement ranging from ± 15% to ± 35% joint movement on a 12.7 mm wide joint are provided in Table 3 below.

**Table 3. Distance traveled in one cycle, for given amplitudes of joint movement**

| **Distance traveled, mm** | **Amplitude of Joint Movement, %** | | | | |
|---|---|---|---|---|---|
| | ± 15 | ± 20 | ± 25 | ± 30 | ± 35 |
| **d (mm)** | 7.62 | 10.16 | 12.7 | 15.24 | 17.78 |

An approximate rate of daily joint movement is obtained from earlier work done by Karpati [2] that gives a value of 0.001 mm/min. for a 12.7 mm wide building joint. This is based on the assumption that daily cycles of ± 4% amplitude of movement are superimposed on annual cycles of ± 25% amplitude. It is the daily cycles that produce the faster rate of movement, and thus the more strenuous cycling action, and so this is the rate used in all following calculations.

Potential sealant lifetimes (in cycles to mode C failure) at different amplitudes of movement can be calculated by using values of m and C as determined from Table 2. Thus, given the number of cycles to failure and the rate of movement for a corresponding amplitude of movement and travel distance, an estimate of the useful lifetime of the sealant specimen (in minutes) can be determined using equation 3 above. The results of these sealant lifetime calculations, provided in years to failure, are given in Table 4 for products L1, L2 and U02 respectively. These values are compared to those provided for a silicone based product, S01, results of which were given in a previous paper [11].

**Table 4. Estimated joint lifetimes for all sealants analyzed.**

| | **Joint Lifetime (years) for Given Amplitudes of Movement** | | | | |
|---|---|---|---|---|---|
| **Sealant** | **± 15%** | **± 20%** | **± 25%** | **± 30%** | **± 35%** |
| **L1** | 0.35 | 0.3 | 0.3 | 0.2 | 0.2 |
| **L2** | 5.1 | 2.7 | 1.7 | 1.1 | 0.8 |
| **U02** | 12.6 | 4.2 | 1.8 | 0.9 | 0.5 |
| **S01** | 33 | 12.8 | 6.1 | 3.4 | 2.0 |

In general, the smaller the value of m, the less sensitive the joint lifetime of a sealant is to changes in amplitude of movement. Product U02 has a large value of m and shows a sharp decrease in lifetime (12.6 to 0.5 years) over the same change in amplitude of joint movement. Predicted lifetime is also affected by C. Since log N is shifted (up or down) by the amount log C, a larger value of C, and therefore of log C, will also increase the predicted lifetime of the joint.

For sealant U02 (Figure 5), the testing of 104 of each of these sealants enough samples had failed to accurately describe failure trends. Sealants L1 and L2 (Figures 8, 9) were not as extensively tested U02 but all samples were tested to failure and all failures were cohesive. The strength of the linear log N and log $\varepsilon$ relationship for these sealants is a clear indication of the potential usefulness of interpreting fatigue results for sealants in this manner, and with the exception of L1, the estimated lifetimes obtained by use of the regression data are reasonable enough to assume that this is a valid method of sealant performance evaluation.

Figure 10 shows how the log N - log $\varepsilon$ plots of the various sealants compare against each other, and graphically reflect the behavior reported in Table 4; they provide an approximate means of ranking the different polyurethane products. Product U02 appears to have the greatest durability at higher strain amplitudes (i.e. lower values of log $\varepsilon$), ahead of L2 and L1, but below a log $\varepsilon$ value of -0.6 (approx. ± 25% amplitude); however, of the three products evaluated, L2 seems to be the most resistant to cyclic fatigue.

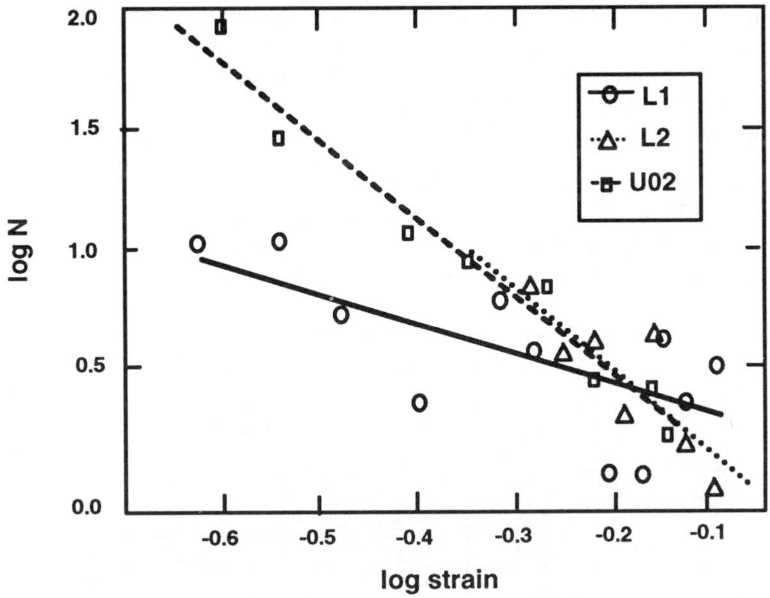


**Figure 10.** Log N (number of cycles to Mode C failure) as a function of log strain (±% amplitude of movement) for products U01, L1 and L2.

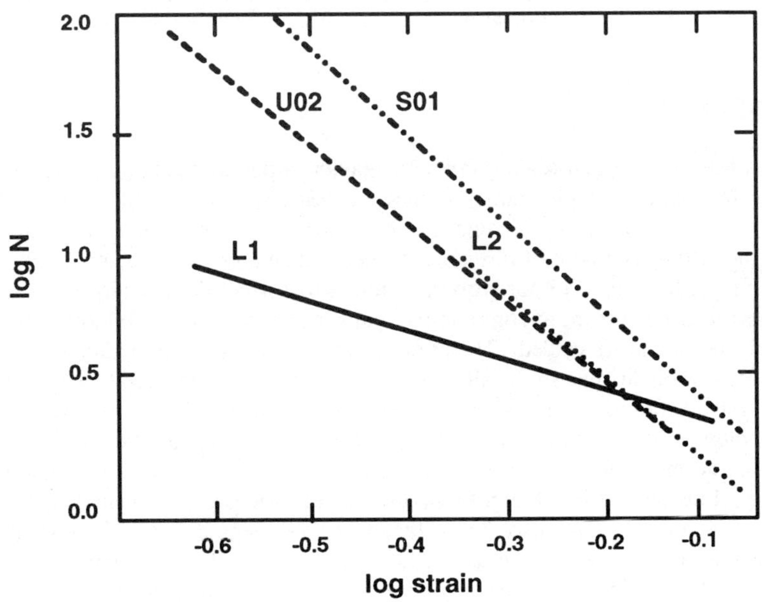


**Figure 11.** Log N (number of cycles to Mode C failure) as a function of log strain (± % amplitude of movement) for products U02, L1, L2 and S01.

Figure 11 contrasts the fatigue behavior of S01, and the polyurethane products, U02, L1 and L2. Owing to its steeper slope, product U02 is slightly more sensitive to changes in the strain amplitude than S01 which, as discussed above, exhibits almost identical behavior. Product S01 also has a longer expected fatigue lifetime since its regression lines are shifted upwards from the U02 log N - log strain regression line.

Typically, silicone sealants are said to be useful up to amplitudes of movement of joint width of ± 50%, while polyurethanes can withstand movements of ± 25%. Ranking the sealants according to their ability to meet their expected performance level gives the results displayed in Table 5.

**Table 5. Ranking of sealants by their ability to meet the expected movement rating of their class.**

| Sealant | Movement Rating | Lifetime (years) |
|---|---|---|
| **U02** | ± 25% | 1.8 |
| **L2** | ± 25% | 1.6 |
| **S01** | ± 50% | 0.62 |
| **L1** | ± 25% | 0.26 |

It is interesting to note that while the silicone S01 outperforms the polyurethanes U02 and L2 at lower amplitude, they do not perform as well at the higher amplitudes that they are supposed to be able to withstand.

## CONCLUSIONS

It has been shown that sealant fatigue behavior in polyurethane products can be analyzed using the same principles that Bartenev models elastomeric fatigue [10]. These results reflect those obtained for silicone sealants reported earlier [8] and essentially demonstrate that the fatigue behavior of different sealants can be compared quantitatively through joint lifetime predictions. Further work is required to perfect the technique as the reliability of the results is dependent not only on the number of specimens tested but also the manner in which specimens are cycled. The effect of varying rate and temperature of testing has not been reviewed in this study although these are factors that will necessarily affect the outcome of fatigue test results. Finally it should also be considered that cycles of movement in joints found in the field often follow periods of inactivity in which there is little or no movement. Hence, products are often in a compressed or extended state for long periods of time in relation to the time spent in actual movement and thus, the type of cyclic movement may also be a factor in the degradation of these products. In consideration of these technicalities, the degree to which results can be affected in tests conducted either in continuous or intermittent movement also needs to be assessed.

## REFERENCES

[1] Karpati, K.K., "Weathering of Silicone Sealant on a Strain-Cycling Exposure Rack'", Adhesives Age, Vol. 23, No. 22, 1980, p.41-47.

[2] Karpati, K.K., "Investigation of the factors influencing the outdoor performance of two-part polysulfide sealants", Journal of Coatings Technology, Vol. 56, No. 719, 1984, p.57-32.

[3] Karpati, K.K., "Testing Polysulfide Sealant Deformation on Vises", Adhesive Age, Vol. 28, No. 5, 1985, pp. 18-22.

[4] Karpati, K.K., "Laboratory Fatigue Testing of a Two-Part Polysulfide Sealant Correlated to Outdoor Performance", Durability of Building Materials, Vol. 5, No. 1, 1987, p. 35-51.

[5] Karpati, K.K., "Exposure Evaluation of Sealants with Low Movement Capability", Adhesive Age, Vol. 31, No. 5, 1988, pp. 20-24.

[6] Karpati, K.K., "Performance of Polyurethane Sealants on a Strain-Cycling Exposure Rack", Materials and Structures, Vol. 22, No. 127, 1989, pp. 60-63.

[7] Karpati, K.K., Solvason, K.R., Sereda, P.J., "Weathering rack for sealants", Journal of Coatings Technology, Vol. 49, No. 626, 1977, p.44-47.

[8] Karpati, K.K., "Device for weathering sealants undergoing cyclic movements". Journal of Coatings Technology, Vol. 50, No. 641, p27-30, June 1978.

[9] Lacasse, M.A., Functional Properties of Polyurethane Based Sealants Blended with Polymeric Modifiers, Ph.D. thesis, Centre for Building Studies, Concordia University, Montreal, 1991.

[10] Bartenev, G.M. and Zuyev, Yu. S., Strength and Failure of Visco-elastic Materials, Pergamon Press Ltd., 1968.

[11] Lacasse, M.A., Bryce , J.E., Margeson , J.C., "Evaluation of Cyclic Fatigue as a Means of Assessing the Performance of Construction Joint Sealants: Silicone Sealants", Science and Technology of Building Seals, Sealants, Glazing, and Waterproofing: Fourth Volume, ASTM STP 1243, David H. Nicastro, Ed., American Society of Testing and Materials, Philadelphia, 1995

Fiorillo, Anthony R.[1]

# EVALUATION OF VARIOUS GENERIC TYPES OF BUILDING SEALANTS AGAINST ASTM C-920, STANDARD SPECIFICATION FOR ELASTOMERIC JOINT SEALANTS

**REFERENCE:** Fiorillo, A. R., **''Evaluation of Various Generic Types of Building Sealants Against ASTM C-920, Standard Specification For Elastomeric Joint Sealants''**, Science and Technology of Building Seals, Sealants, Glazing and Waterproofing: Fifth Volume, ASTM STP 1271, Michael A. Lacasse, Ed., American Society For Testing and Materials, 1996.

**ABSTRACT:** A number of sealant manufacturers suggest that although they supply primers for their sealants, in many cases the primers may not be necessary. They usually suggest running a test on the substrates the sealants are to be used with to determine if the primers are needed. This paper will discuss the results of testing several commercially available polysulfide, silicone and polyurethane sealants against ASTM C-920, Standard Specification For Elastomeric Joint Sealants. Mortar, plate glass and anodized aluminum were used as the substrates. Where manufacturers supplied primers, the sealant was tested with and without their primer. Where the manufacturer did not recommend primers, testing was only done without a primer.

**KEYWORDS:** Polyurethane, Polysulfide, Sealant, Silicone

When an architect specifies a sealant for a particular application, he will usually specify the particular sealant he wants to be used but he will also allow an equivalent sealant to be substituted, giving the contractor some freedom to choose the sealant he will use.

When the contractor is going to apply the sealant, if he has a choice he will usually pick one he is familiar with. In most cases the sealant the contractor applies will perform satisfactorily. But, occasionally, there may

[1] Group Leader, Liquid Polymers Technical Service and Applications, Morton International, Inc., Woodstock, IL 60098.

be some special requirement of the sealant that may not make his choice the best one for the application.

There are a number of high quality silicone, urethane and polysulfide sealants the contractor can choose from. Today even some acrylic sealants are advertised as having properties comparable to the silicones, polyurethanes and polysulfides.

All sealant manufacturers have technical data sheets that give test results of their sealants, so a contractor can try comparing technical data sheets in an effort to choose a sealant. This can be confusing because, although manufacturers usually specify how they did their testing, not all companies will do it the same way.

Even in cases where the testing is done according to the same specification, testing conditions at one lab will not be identical to testing conditions at another lab, and this will also give confusing results.

Often the difference between a sealant working successfully or not depends on whether a primer was used with the sealant. A number of sealant manufacturers say that for most applications primers may not be necessary with their sealants. They usually will suggest running a test on the substrates the sealant is to be used with to determine if primers are needed. Since occasionally a contractor may feel it is not convenient to run a test to determine if a primer is needed, where primers were used, the sealants were also tested without the primer to compare the results.

When a contractor is planning to apply a building sealant and he has been given the freedom of choosing an equivalent sealant, if he is not familiar with all the various generic types available, he will have to try to determine what an equivalent sealant is.

Few studies have been done comparing the various generic types of sealants so usually when a contractor has to choose a sealant, he will choose one he has experience with, afraid that if he were to switch to another generic type it may not perform as well.

ASTM is publishing work done by Chuck Parise which gives the most important characteristics of each generic type sealant. This will be a helpful guide for the architect in specifying sealants for particular applications and for the contractor in choosing an equivalent sealant.

A study done by one lab comparing various generic type sealants, tested under identical conditions, will help a contractor to make a more accurate comparison between sealants. But to be completely helpful, a series of studies should be done. These studies might include, testing against applicable specifications, testing against all materials commonly used in building construction and testing under various environmental conditions.

In this study ten different commercially available building sealants were tested by the same lab against ASTM C-920. The study included two silicone, three polysulfide and five urethane sealants. One of the polysulfide and one of the urethane sealants were multi-component sealants. The other eight were all single component sealants.

Mortar, plate glass and anodized aluminum were used as the substrates. Where manufacturers supplied primers, the sealant was tested with and without their primer. Where the manufacturer did not recommend primers testing was only done without a primer.

Most of the sealants met all the requirements of the specification except for the type of failure in the peel adhesion test. One of the polysulfide sealants was not completely cured so it did not meet the minimum hardness requirements and three of the urethane sealants exceeded the maximum hardness after heat aging.

Only one of the ten sealants failed to meet the weight loss requirements. This same sealant is the only one that did not meet the cyclic movement testing requirements. This sealant also developed a large number of channel like voids in the sealant which can serve as paths for air and moisture to penetrate through the sealant, defeating the purpose of the sealant.

A number of the sealants failed the peel adhesion test. Here, most of the failures were in the amount of bond loss. In most cases where there was excessive bond loss, the peel strength value was high, but there were several cases where the strength was quite low and probably would have led to sealant failure in an actual application.

Table I

COMPARISON OF VARIOUS GENERIC TYPE SEALANTS WHEN TESTED AGAINST ASTM C-920, STANDARD SPECIFICATION FOR ELASTOMERIC JOINT SEALANTS (Primer used if recommended)

| Sealant | 1 | 2 | 3 | 4 | 5 | 6 | 7 | 8 | 9 | 10 |
|---|---|---|---|---|---|---|---|---|---|---|
| Type | PS | PS | PS | PU | PU | PU | PU | PU | S | S |
| Indentation Hardness of Elastomeric-Type Sealants by Means of a Durometer (ASTM C-661) | | | | | | | | | | |
| Initial | 23 | 25 | 10 | 17 | 25 | 35 | 40 | 40 | 18 | 15 |
| Heat aged | 25 | 28 | 20 | 19 | 30 | 55 | 53 | 55 | 26 | 28 |
| Effects of Heat Aging on Weight Loss, Cracking and Chalking of Elastomeric Sealants (ASTM C 792) | | | | | | | | | | |
| Wt. Loss, % | 2.3 | 2.6 | 4.8 | 6.1 | 8.0 | 8.8 | 4.9 | 11.4 | 1.3 | 3.5 |
| Staining and Color Change of Single or Multicomponent Joint Sealants (ASTM C 510) | | | | | | | | | | |
| | No | No | No | No | No | No | No | Yes | No | No |
| Effects of Accelerated Weathering on Elastomeric Joint Sealants (ASTM C 793) | | | | | | | | | | |
| Weather | 0 | 0 | 0 | 0 | 0 | 0 | 0 | 0 | 0 | 0 |
| + bending | 0 | 0 | 0 | 0 | 0 | 0 | 0 | 0 | 0 | 0 |

* Sealant not cured in center of sample.

Table I (continued)

COMPARISON OF VARIOUS GENERIC TYPE SEALANTS WHEN TESTED AGAINST ASTM C-920 (Primer used if recommended)

| Sealant | 1 | 2 | 3 | 4 | 5 | 6 | 7 | 8 | 9 | 10 |
|---|---|---|---|---|---|---|---|---|---|---|
| Type | PS | PS | PS | PU | PU | PU | PU | PU | S | S |
| Adhesion and Cohesion of Elastomeric Joint Sealants Under Cyclic Movement (ASTM C 719-Tested ±25%) (Bond loss in $cm^2$) | | | | | | | | | | |
| Mortar | 0 | <3 | >9.7* | <3 | 0 | 0 | 6 | >9.7 | 0 | 0 |
| Aluminum | 0 | <3 | >9.7* | <3 | 0 | 0 | <3 | >9.7 | 0 | 0 |
| Glass | 0 | <3 | >9.7* | <3 | 0 | 0 | 0 | >9.7 | 0 | 0 |
| Adhesion-in-Peel of Elastomeric Sealants (ASTM C-794) (Values in Kg) | | | | | | | | | | |
| Mortar | 7.2<br>100%<br>coh | 8.3<br>100%<br>coh | 1.9* | 13.4<br>50%<br>coh | 16.3<br>90%<br>coh | 15.3<br>100%<br>coh | 19.7<br>10%<br>coh | 7.2<br>0%<br>coh | 0.4<br>0%<br>coh | 18.6<br>0%<br>coh |
| Aluminum | 5.9<br>100%<br>coh | 17.4<br>100%<br>coh | 2.5* | 12.0<br>100%<br>coh | 15.0<br>80%<br>coh | 13.1<br>50%<br>coh | 25.3<br>0%<br>coh | 9.8<br>20%<br>coh | 17.4<br>5%<br>coh | 1.6<br>0%<br>coh |
| Glass | 6.7<br>100%<br>coh | 15.5<br>100%<br>coh | 3.2* | 13.3<br>100%<br>coh | 16.8<br>80%<br>coh | 10.3<br>50%<br>coh | 24.0<br>0%<br>coh | 8.1<br>85%<br>coh | 17.1<br>5%<br>coh | 19.0<br>50%<br>coh |
| After UV (Through Glass) | 5.4<br>100%<br>coh | 20.7<br>100%<br>coh | Not run | 4.26<br>0%<br>coh | 15.9<br>50%<br>coh | 4.7<br>60%<br>coh | 23.6<br>50%<br>coh | 12.4<br>50%<br>coh | Not run | Not run |

* Sealant not cured in center of sample.

Table II

| COMPARISON OF VARIOUS GENERIC SEALANTS WITHOUT THE USE OF PRIMERS | | | | | | | | | | |
|---|---|---|---|---|---|---|---|---|---|---|
| Sealant | 1 | 2 | 3 | 4 | 5 | 6 | 7 | 8 | 9 | 10 |
| Type | PS | PS | PS | PU | PU | PU | PU | PU | S | S |
| Adhesion and Cohesion of Elastomeric Joint Sealants Under Cyclic Movement (ASTM C 719-Tested ±25%) (Bond loss in $cm^2$) | | | | | | | | | | |
| Mortar | 0 | <3 | Not Run | <3 | <3 | 0 | 9.7 | 9.7 | 9.7 | 0 |
| Aluminum | 0 | <3 | 2.5 | <3 | 6 | 0 | 6.2 | 9.7 | 0 | 0 |
| Glass | 0 | <3 | 3.2 | <3 | <3 | 0 | 0 | 6 | 0 | 0 |
| Adhesion-in-Peel of Elastomeric Joint Sealants (ASTM C-794) (Values in Kg) | | | | | | | | | | |
| Mortar | 7.2<br>100%<br>coh | 8.3<br>100%<br>coh | Not<br>run | 9.9<br>80%<br>coh | 15.4<br>0%<br>coh | 14.0<br>100%<br>coh | 3.8<br>0%<br>coh | 7.3<br>0%<br>coh | 1.4<br>0%<br>coh | 18.6<br>0%<br>coh |
| Aluminum | 5.9<br>100%<br>coh | 17.4<br>100%<br>coh | 2.5* | 13.3<br>90%<br>coh | 15.9<br>60%<br>coh | 18.5<br>60%<br>coh | 19.8<br>20%<br>coh | 7.9<br>20%<br>coh | 17.4<br>5%<br>coh | 1.6<br>0%<br>coh |
| Glass | 6.7<br>100%<br>coh | 15.5<br>100%<br>coh | 3.2* | 11.0<br>100%<br>coh | 20.4<br>10%<br>coh | 25.0<br>75%<br>coh | 24.0<br>0%<br>coh | 8.0<br>85%<br>coh | 17.1<br>5%<br>coh | 19.0<br>50%<br>coh |
| After UV (Through Glass) | 5.4<br>100%<br>coh | 20.7<br>100%<br>coh | Not<br>run | 4.1<br>0%<br>coh | 13.1<br>0%<br>coh | 29.5<br>0%<br>coh | 21.8<br>50%<br>coh | 10.0<br>50%<br>coh | Not<br>run | Not<br>run |

* Sealant not cured in center of sample.

#2 and #7 are multi-component sealants. All others are single component sealants.

All sealants tested are commercially available sealants, most of which have been used in building construction for years. Despite this fact, these test results show that some performed much better than others in this particular evaluation.

The results also show:

1. Some single component polysulfide sealants are extremely slow curing.

2. Since a number of the sealants included in this study did not meet the maximum bond loss of C-794, even though they have a history of success in the field, it might be worth while for ASTM to consider modifying C-794 to give the option of either meeting the specification as it is now written or to increase the permissible bond loss as the peel strength goes up.

3. The loss of strength shown by some of the urethane sealants after exposure to UV through glass indicates that perhaps ASTM should also consider making step 7.1.12 (peel adhesion after exposure to UV through glass) of C-794 mandatory.

4. Although some of the sealants recommend the use of primers, in a number of cases test results did not show improvement with the primer, especially on non-porous substrates.

Putman, William J.[1], Pekara, David[1], and McGreer, Matthew[1]

**A REVIEW OF RECENT DEVELOPMENTS AND IMPROVEMENTS IN ACCELERATED OUTDOOR AND INDOOR XENON WEATHERING DEVICES AND METHODOLOGIES**

**REFERENCE:** Putman, William J., Pekara, David, and McGreer, Matthew, ''**A Review of Recent Developments and Improvements in Accelerated Outdoor and Indoor Xenon Weathering Devices and Methodologies**'', Science and Technology of Building Seals, Sealants, Glazing, and Waterproofing: Fifth Volume, ASTM STP 1271, 1996.

**ABSTRACT:**

Accelerated outdoor weathering has a history which extends back at least 50 years. The advantage of this type of testing is that the sun's natural solar energy spectrum is used to irradiate the specimens under test. Since this is an outdoor method, variations in test conditions occur due to the natural seasonal weather differences. Recent improvements in fresnel concentrator accelerated weathering devices such as day and nighttime temperature control, and new moisture cycles are described that minimize seasonal effects.

Since the development of artificial tests in the early 1900's, many types of artificial light sources have been introduced. Improvements in xenon accelerated weathering chambers have enabled development of new test methods utilizing filters that provide a very close approximation of sunlight especially when compared to fluorescent and carbon arc light sources. This paper will also include a description of an advanced air-cooled xenon arc test apparatus for the accelerated exposure of materials.

**KEYWORDS:** outdoor accelerated weathering, fresnel concentrator, xenon weathering, ultra violet radiation

## OUTDOOR ACCELERATED WEATHERING

The first commercial use of follow-the-sun racks for outdoor accelerated weathering was in the late 1950's by Caryl, et al.[1] During the 1960's, mirrors were attached to the equatorial rack for greater acceleration. These mirrors focused sunlight onto specimens which were mounted onto an air-cooled specimen plane. These accelerated weathering devices were designated by DSET as EMMAQUA®, Equatorial Mount with Mirrors for Acceleration plus Aqua (water). The device can be operated without spray, designated as EMMA®, or with sprays as either EMMAQUA® or EMMAQUA® Nighttime Wetting (NTW). A complete description of the EMMAQUA® device is provided in Reference 2. This apparatus is referenced in numerous national and international standards and materials specifications, such as ASTM G 90, D 4141, D 4364,; SAE J576 and J1961, and ISO 877. The primary technical advantage of the EMMAQUA® test method is that it uses the sun's natural solar spectrum. This has

[1] Manager and supervisors of Testing, DSET Laboratories, 45601 N. 47th Avenue, Phoenix, Arizona 85027-7042.

allowed the method to correlate quite well to southern Florida exposures for a wide variety of materials.[3-9]

In 1991, the EMMAQUA® device was redesigned by DSET as shown in Figure 1. The newly designed machine was designated EMMAQUA®+ and incorporated features such as dual axis tracking and an improved water spray system. The standard ten mirrors which focus the direct beam of the sun onto the targetboard, the blower and air tunnel for specimen cooling did not change. The EMMAQUA®+ machine can also be inverted in order to position test specimens 5° from horizontal during nighttime moisture cycles. This specimen position for nighttime wetting may better simulate southern Florida moisture conditions for hydrophobic materials.

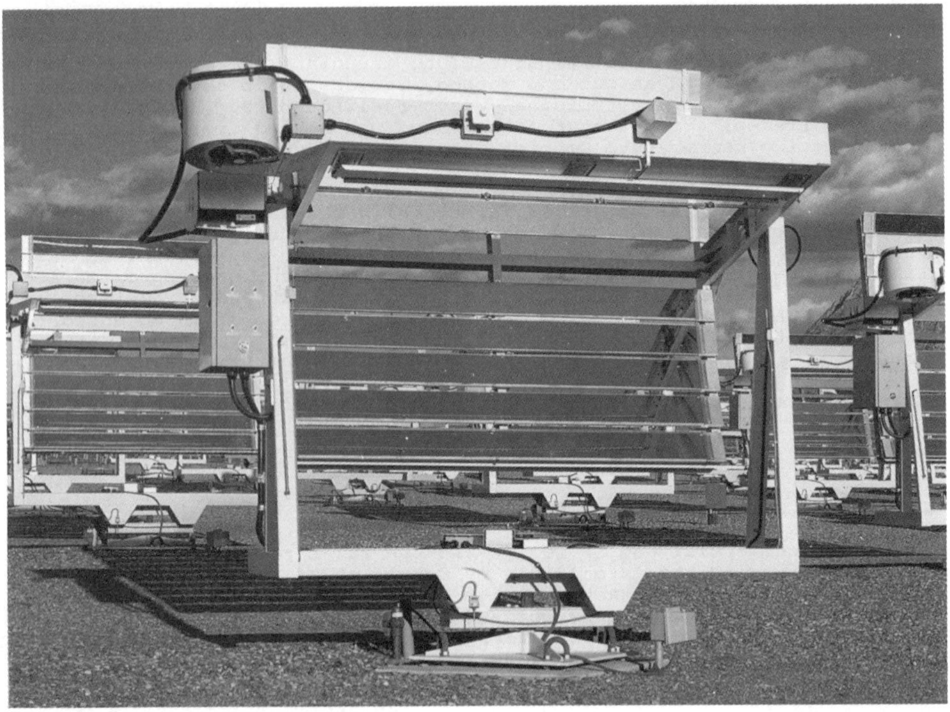

Figure 1. EMMAQUA®+ Device

**EMMAQUA®+ TEMPERATURE CONTROL**

The most significant feature added to the EMMAQUA®+ device is test specimen temperature control. Sample temperature is set by combining three methods of control with a microprocessor. The primary features of this system are:

- Backside sample heating
- Sample cooling air velocity control
- Defocusing of the machine

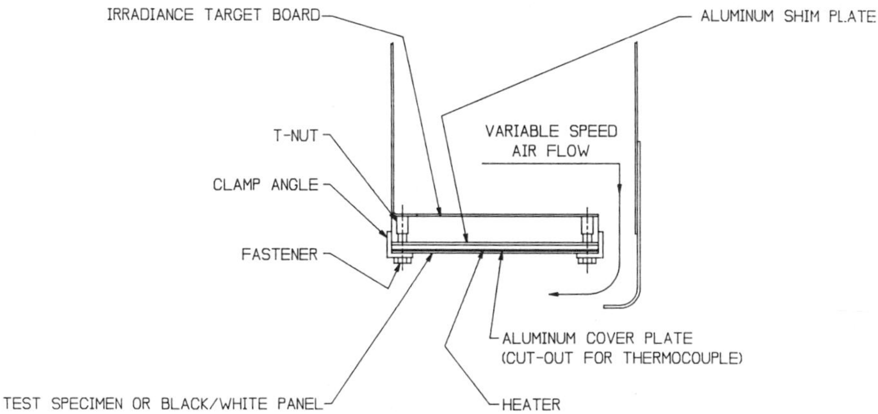


Figure 2. Schematic of EMMAQUA®+ Specimen Temperature Control Set-up

Temperatures of black and white panels, of known solar absorptance, are continuously measured with type T thermocouples and mounted on the machine adjacent to the test specimens. The black and white panel temperatures are monitored with a programmable logic controller/data logger. A cross-section diagram of the temperature control system is shown in Figure 2. The algorithms contained within the program control the heater output and cooling air speed to maintain black and white panel temperatures at preset values. Black and white panel temperatures are monitored continuously for both day and night cycles. During the day, the black panel temperature is set 15°C greater than the white panel temperature, which appears to be the average difference between black and white panels in Southern Florida during periods of high solar irradiance. At night, without solar irradiance, heaters maintain the black and white panel temperature to the same preset values. The defocus control is an additional safety feature that keeps the sample temperature within specification should there be an equipment failure or if there is an extremely high ambient temperature that could cause a sample overtemperature condition. Figure 3 depicts the desired EMMAQUA®+ temperature control and Figure 4 depicts the typical black and white panel temperature control.

In association with the ability to control specimen temperature, deionized water spray cycles occur during daily EMMAQUA®+ machine run time every hour to provide for specimen wet time and temperature cycling. Each spray period lasts eight minutes. For the first half of the spray period, the machine is driven out of focus to induce a thermal shock effect to specimens. The machines are then allowed to refocus during the second half of the spray cycle to quickly bring specimen temperatures up to normal conditions. Table 1 presents the year-round temperature operating range of the EMMAQUA®+ and available moisture cycles. The variety of moisture cycles allows test users to choose the moisture type most suitable for their material.

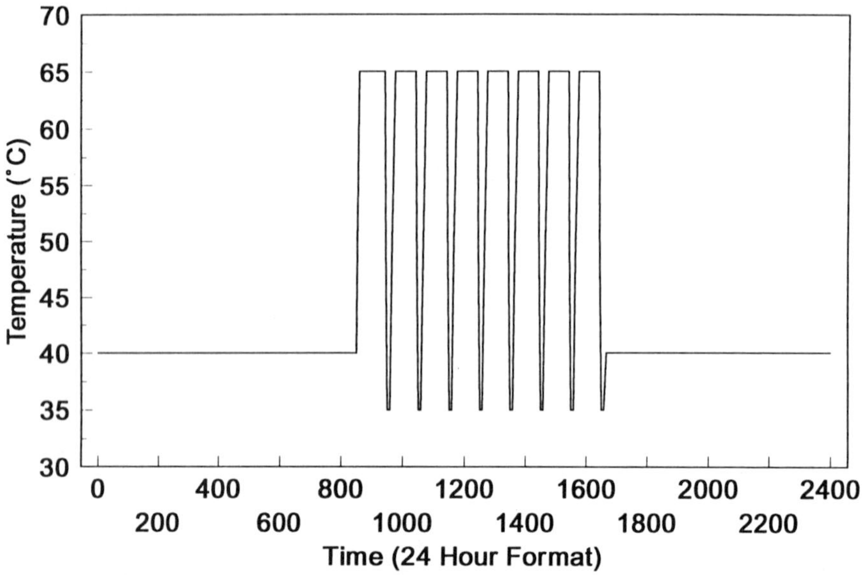


Figure 3. Desired 24 Hour Black Panel Temperatures for EMMAQUA®+ ($T_{BP}$= 65°, $T_{WP}$ = 50°C, Nighttime Temperature = 40°C and hourly daytime sprays)

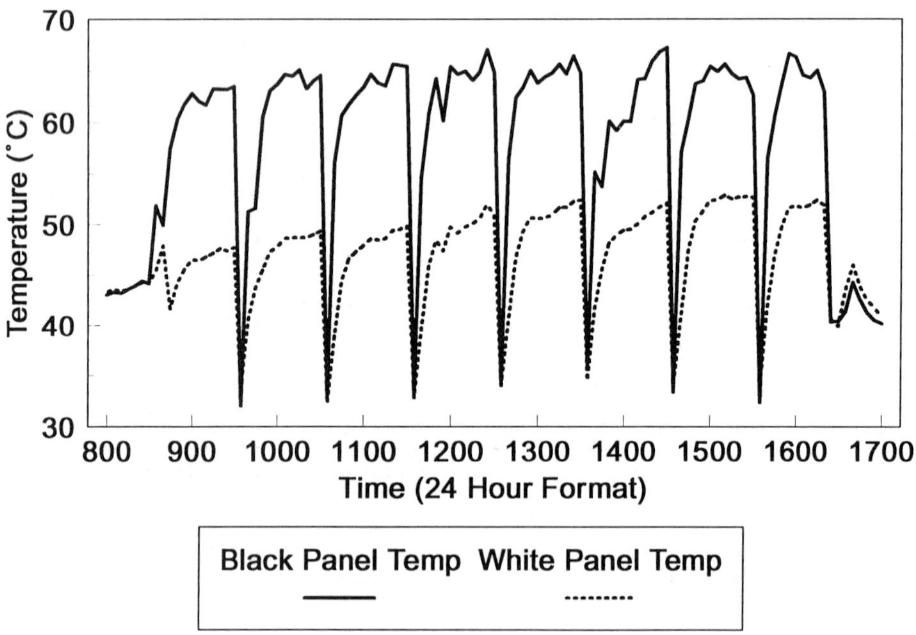


Figure 4. Typical Black and White Panel Temperature Control ($T_{BP}$ = 65°, $T_{WP}$ = 50°C)

| Parameter | | | | | |
|---|---|---|---|---|---|
| Nighttime Moisture Types | Soak-Freeze-Thaw | Soak | Conden-sation | Extended NTW | 5° Dew |
| Nighttime Sample Temperatures | 1 hr. @ 40°C Soak, 12 hrs at -18°C, then thaw at 25°C | 20°C or 40°C | 40°C or 50°C | 30°C, 40°C or 50°C | 30°C, 40°C or 50°C |
| Daytime Black Panel Temperature Range | 65°C to 80°C for all Moisture Cycles | | | | |
| Daytime White Panel Temperature Range | 50°C to 60°C for all moisture Cycles (Must be at least 15°C lower than Black Panel Set Point) | | | | |
| Daytime Moisture | 8 Minutes Sprays Every Hour for all Cycles | | | | |

Table 1. EMMAQUA®+ Temperature Range and New Moisture Cycles

Artificial Weathering Devices

Artificial weathering is the exposure of materials to laboratory controlled conditions of light, heat and moisture. The goal is to achieve accelerated degradation of materials using radiation that is more intense than natural solar radiation.

The spectrum produced by an artificial light source is extremely important in assuring that results correlate to real-time outdoor exposures. Since the development of artificial tests in the early 1900's, many types of artificial light sources have been introduced. The three most common are the carbon arc, fluorescent, and xenon-arc lamps.

Carbon arc devices generally use one or more carbon burners either open or encased in a borosilicate glass cover that acts as a filter for low wavelength radiation. The open flame carbon arc spectral distribution exhibits a significant amount of ultraviolet energy below 300 nm. Radiation in the 300 nm to 350 nm region and in certain wavebands between 350nm and 400nm is deficient compared to natural sunlight. Materials sensitive to these regions of the UV spectrum generally will not exhibit degradation consistent with what would be seen under real-time outdoor conditions.

Fluorescent devices use one or more fluorescent lamps with special phosphors to produce ultraviolet radiations. The most commonly used lamp emits UV energy between approximately 280 and 350nm with the peak in the 310-315nm range. Recent advances in these lamps have provided improved correlation to outdoor solar spectral distribution.

Xenon-arc devices consist of one to three water or air cooled lamps. The pure xenon spectrum is filtered to produce a spectral distribution that closely matches natural sunlight. As seen in Figure 5, properly filtered xenon lamps produce a spectral distribution that approximates sunlight more closely than either the carbon-arc or fluorescent lamp. This is a primary reason for the increase in the use of xenon devices.

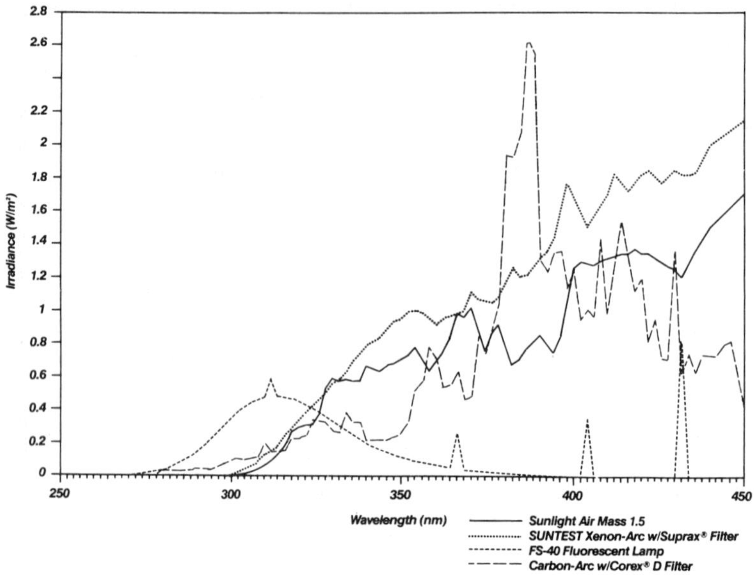


Figure 5. Comparison of Lamp Spectra

Even though xenon devices have the best simulation of the natural ultraviolet solar spectrum, complaints that are sometimes voiced about xenon weathering chambers are that they are too noisy, hard to maintain, hard to operate and hard to calibrate.

The recently introduced Xenotest® Beta LM unit has many design features intended to improve the overall ease of operation of a Xenon weathering chamber. Figure 6 shows the solar spectral distribution for the Beta as compared to natural sunlight and Figure 7 depicts the operating range of the Beta.[10] These figures show that the Beta's operating parameters are consistent with the requirements of current xenon weathering standards. The advantages the Beta has over older type xenon weathering devices are:

- Quiet operation
- Digital measurement and control of irradiance, chamber temperature, black standard temperature and relative humidity
- Computer linkable microprocessor that controls and displays test parameters
- Diagnostic software for trouble shooting
- Sample plane irradiance and black standard temperature measurement and control
- Xenosensiv measurement and calibration system
- No water cooling of the lamps or filters because of an improved filter
- Front and/or backside spray

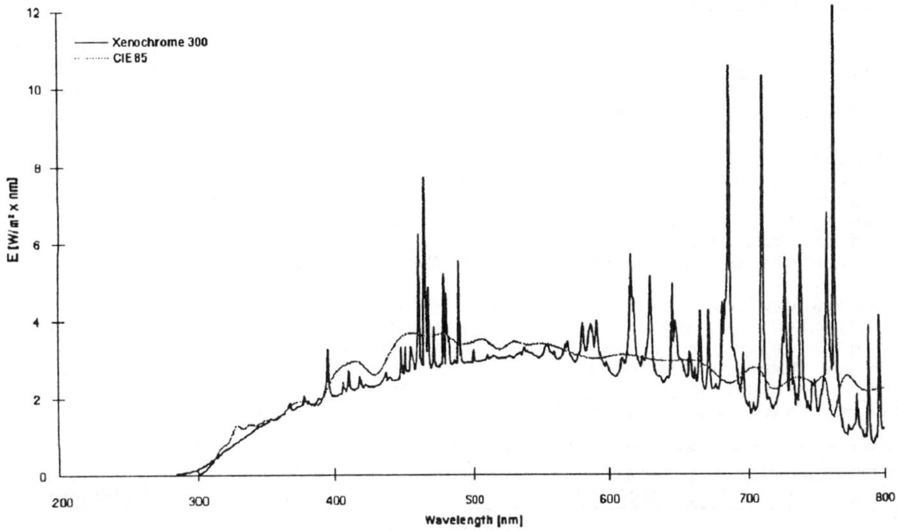


Figure 6. Xenotest® Beta LM Spectrum with Xenochrome 300/Suprax Filter Versus Natural Sunlight

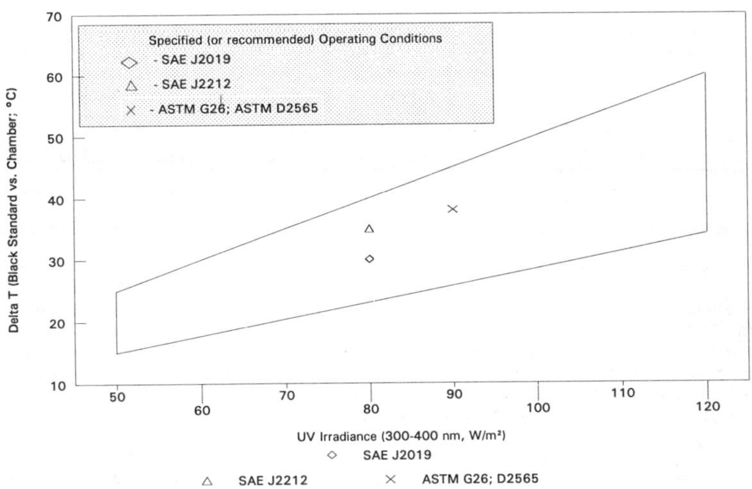


Figure 7. Xenotest® Beta LM Operating Ranges

Consequently, the Beta can meet the parameters of existing standards while incorporating features that enable it to be quieter, and easier to maintain, operate and calibrate when compared to older design xenon weathering devices.

**TEMPERATURE COMPARISON BETWEEN EXPOSURE TYPES**

When developing accelerated test cycles, defining the real-time climate is critical since accelerated weathering cycles usually incorporate the extremes of an environment especially in the area of sample temperature and temperature cycling.

Methods of temperature control for indoor and outdoor accelerated weathering devices have been developed to: 1) better simulate real-time exposure temperatures, 2) keep specimen temperature at near-maximum levels, and 3) maintain proper temperature relationships of difference specimen colors. Testing was recently conducted to verify the temperature control methods of the EMMAQUA®+ and Xenotest® Beta LM.

Temperature measurements were performed under four separate exposure conditions: 1) Arizona real-time exposure, 2) southern Florida real time exposure, 3) temperature controlled outdoor accelerated weathering device (EMMAQUA®+) and 4) indoor xenon arc weathering device (Beta). Both plastic and paint specimens of varying colors, solar absorptance values, and compositions were tested. (See Table 2).

| Specimen Color | Material Type | Material Description | Solar Absorptance |
|---|---|---|---|
| White | Plastic | MDPE (2 mm) | 0.25 |
| Green | Plastic | PVC (1.2 mm) | 0.49 |
| Gray | Plastic | PVC (1.1 mm) | 0.48 |
| Black | Plastic | Polypropylene-EDPM (2 mm) | 0.95 |
| White | Paint | Polyurethane | 0.19 |
| Yellow | Paint | Coil Coating | 0.54 |
| Blue #1 | Paint | Acrylic Clear Coat Basecoat | 0.71 |
| Blue #2 | Paint | Polyester Coil Coating | 0.77 |
| Brown | Paint | Silicone Polyester Coil Coating | 0.55 |

Table 2. Specimen Description

All measurements were taken with 30 AWG Type "T" thermocouples. Thermocouples on thin specimens (less than 2 mm) were mounted on the surface away from the light source. If the thickness of the specimen allowed, the thermocouple was secured into a small hole drilled into the side. All measurements were recorded with a digital microprocessor thermometer or data acquisition system.

Real-time outdoor temperature measurements were taken on clear, hot days where near-maximum ambient temperatures existed. Specimens were mounted without a backing at latitudinal angles (34°- Arizona, 26°- Florida), except the black panels, which were exposed at 45° south in both Arizona and southern Florida. Maximum temperatures of four specimen types (black plastic, black paint, white plastic, and white paint) for Arizona and Florida exposures are shown in Figure 8.

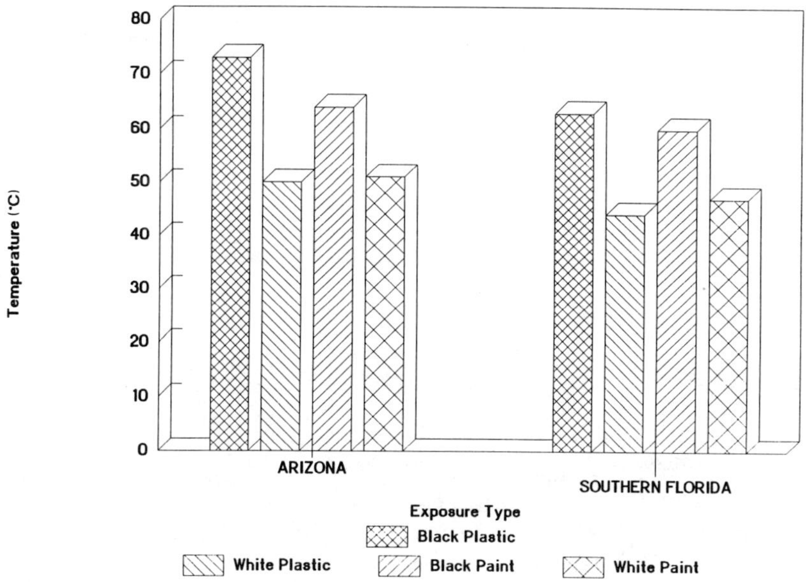


Figure 8. Maximum Black/White Paint/Plastic For Arizona and southern Florida Exposures

Specimens on the outdoor accelerated weathering device were mounted as shown in Figure 2. Paints and plastics were tested separately. Paint panel tests utilized the black and white panel temperatures for control. Black and white plastic temperatures controlled the plastics test.

Tests in the indoor xenon weathering device used the standard SAE J2019 specification for test parameters. Irradiance was controlled at 80 $W/m^2$. Relative humidity was controlled at 50%. Chamber temperature and black standard temperature were maintained at 47°C and 76°C, respectively. The specimens were mounted on stainless steel plates in the specimen holder.

Analysis of temperature data consisted of comparing the specimen temperature of different colors to the white panel for all four exposure types. Data were compared to examine if proper temperature relationships existed between colors, and if these temperature relationships between exposure types were comparable. Graphs were then generated showing the temperature difference when compared to the white panel. (See Figures 9 and 10).

Dark colored specimens (increasing solar absorptance) had higher temperatures than the white panel, as expected. Differences between panel temperatures were similar for all exposure types. Control of the white panel temperature and black panel temperature maintained a temperature relationship similar to real time exposures.

For the EMMAQUA®+ and Beta, a significant improvement was seen in maintaining colors at or near temperatures that would be experienced in a conventional weathering exposure. Colors between black and white had a tendency to be about 5°C higher than would be predicted from outdoor test data.

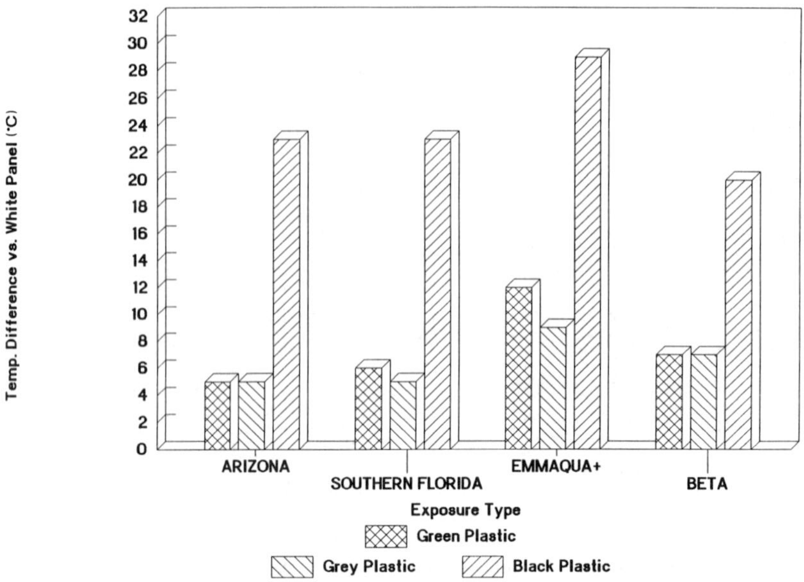


Figure 9. Temperature Differential vs. White Panel for Each Exposure Type (Plastics)

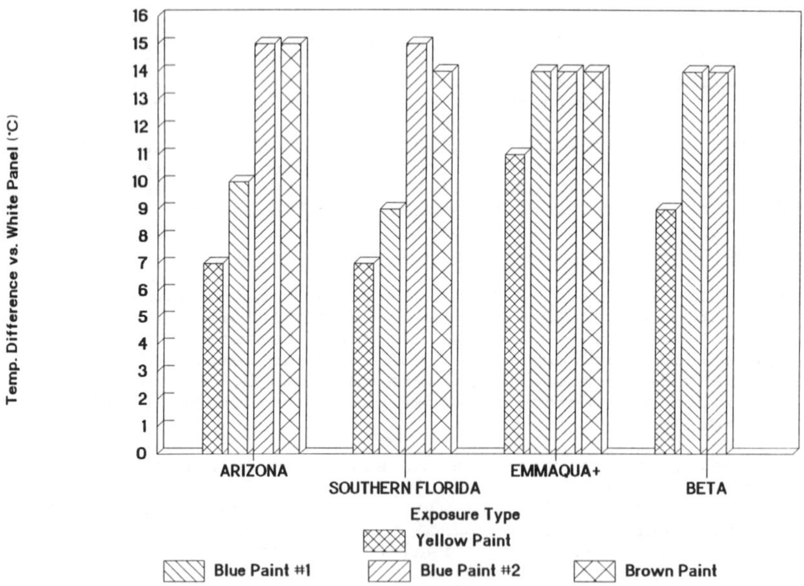


Figure 10. Temperature Differential vs. White Panel for Each Exposure Type (Paints)

## CONCLUSIONS

Recent improvements to the fresnel concentrator accelerated weathering devices have incorporated temperature control to minimize seasonal effects and have resulted in the development of new test cycles. These test cycles closely simulate temperature and moisture parameters of real-time weathering. In addition, these cycles induce thermal shocks and provide a wider range of parameters for different material types. Testing is ongoing to attempt to determine the best suited cycles for these materials.

Improvements in xenon devices now allow for a closer match to natural sunlight, especially in the ultraviolet portion of the spectrum. New xenon devices, such as the Xenotest® Beta LM, have eliminated the need for lamp or filter water cooling through the use of multilayer dielectric type filter systems. These factors combined with a user friendly microprocessor and the Xenosensiv monitoring and calibration system should make this type of xenon device quieter and easier to operate, maintain and calibrate when compared to older design xenon weathering chambers.

Temperature test results for four different exposure types verified indoor and outdoor accelerated weathering devices exposed specimens to similar maximum thermal conditions found in Arizona and Florida environments. It was also determined that proper temperature relationships were maintained for specimens of varying color and composition on both the indoor and outdoor testing devices.

## REFERENCES

[1] Caryl, C.R. and Helmick, W.E., U.S. Patent No. 2,945,417, 19 July 1960.

[2] Robbins, J. S. III, "A Review of Recent Developments in the Use of ASTM Standard Practice G 90 for the Testing of Nonmetallic Materials," Accelerated and Outdoor Durability Testing of Organic Materials, ASTM STP 1202. Warren D. Ketola and Douglas Grossman, Eds., American Society for Testing and Materials, Philadelphia, PA 1994

[3] Zerlaut, G.A. and Robbins, J.S. Jr., "Accelerated Outdoor Exposure Testing of Coil Coatings by the EMMAQUA® Test Method," Presentation at European Coil Coaters Association Meeting, Brussels, November 1984.

[4] Higgins Dr., Richard J., "Powder Coatings" Metal Architecture.

[5] National Coil Coaters Association, "Accelerated Weathering Devices, Comparison and Evaluation," Technical Bulletin III-10, October, 1978.

[6] Sacher, Robert E., "Proceedings of the TTCP-3 Critical Review Techniques for the Characterization of Polymeric Materials," Army Materials and Mechanics Research Center, January 1977, Watertown, Massachusetts

[7] Morse, M.P., "Accelerated Outdoor Weathering Tests for Evaluating the Durability of Coatings," Symposium on Permanence of Organic Coatings, ASTM STP 781, G.G. Schurr, Ed., American Society for Testing and Materials, Philadelphia, PA 1981

[8] Bauer, D.R. et. al., "Evaluation of Accelerated Weathering Tests for a Polyester-Urethane Coating Using Photoacoustic Infrared Spectroscopy," Journal of Coatings Technology, Volume 59, No. 755, pp. 103-109, Federation of Societies for Coatings Technology, Philadelphia, PA.

[9] Wineburg, J.P. "Automotive Coatings and Stabilizers," November 12, 1993, Philadelphia, PA 19146

[10] Boxhammer, Dr. Jörg, "Current Status of Light and Weatherfastness Standards - new equipment technologies, operating procedures and application of standard reference materials," 2nd International Symposium on Weatherability (2nd ISW), Materials Life Society, Japan, September 1994.